HEAT

Books by M. Nelkon

Light and Sound (SI) Heinemann
Mechanics and Properties of Matter (SI) Heinemann
Advanced Level Practical Physics (with J. Ogborn) *(SI)* Heinemann
Electricity (Advanced level, SI) Edward Arnold
Fundamentals of Physics (O-level, SI) Chatto & Windus
Exercises in Ordinary Level Physics (SI) Chatto & Windus
C.S.E. Physics Chatto & Windus
Revision Book in C.S.E. Physics Chatto & Windus
SI Units: An Introduction for A-level Chatto & Windus

HEAT

A textbook for advanced level in SI units

M. Nelkon

M.Sc., F.Inst.P., A.K.C.

formerly Head of the Science Department,
William Ellis School, London

BLACKIE
London & Glasgow

Blackie & Son Limited
5 Fitzhardinge Street
London W.1
Bishopbriggs, Glasgow

Blackie & Son (India) Limited
103/5 Fort Street, Bombay

First published 1949
Second edition 1964
Third edition 1970
Reprinted 1971, 1972, 1973, 1978

Printed in Great Britain by
R. & R. Clark Ltd, Edinburgh

Preface to third edition

In this edition the book has been revised and re-ordered to take account of recent changes in the Advanced level syllabus of the Examining Boards. Among the more important changes are:

(1) the use of SI units throughout the book, in accordance with future requirements at Advanced level examinations;
(2) emphasis of the kinetic theory of gases, molecular explanations of change of state, thermal expansion and conduction;
(3) a briefer treatment of the thermal expansion of solids and liquids, and exclusion of 'mechanical equivalent of heat'.

Exercises now contain more recent examination questions, and revision papers have been added. I am indebted to the following examination boards for kind permission to translate past questions to SI units—the translation is the sole responsibility of the author: London University School Examinations (*L.*), Joint Matriculation Board (*N.*), Oxford and Cambridge Schools Examination Board (*O. and C.*), University of Cambridge Local Examinations Syndicate (*C.*), Oxford Delegacy of Local Examinations (*O.*).

The Preface to the first and second editions said:

"This textbook aims at a clear presentation of the fundamental principles of Heat to an Advanced Level or Intermediate standard, and assumes an Ordinary level knowledge of the subject. Numerically worked examples from past examination papers of the various Boards have been included in the text to illustrate the topics, and the calculus has been used only where necessary.

It is not possible here to comment on more than a few of the points borne in mind in the presentation of the subject. The basic principles of thermometry and pyrometry have been fully discussed; the importance of the mass of the gas in Gas Equations has been emphasized; the first law of thermodynamics has been applied to isothermal and adiabatic changes; and I have added an Introduction to Thermodynamics and an account of Order of Accuracy."

I am grateful to the following for their generous assistance with the book: R. P. T. Hills, St. John's College, Cambridge; J. M. Ogborn, Worcester College of Education; Dr. D. L. Pursey, formerly of King's College, London University; P. Parker, late of The City University, London.

In addition to the examining boards mentioned earlier, I am indebted to the following for their kind permission to reprint questions set in past examinations: Central Welsh Board (*W.*), Cambridge Entrance Scholarships and Exhibitions Examinations Board (*C.S.*).

Contents

1
Temperature and Measurement

Heat and temperature

The molecules of all substances, solid, liquid, or gas, are continually in motion. This is called *thermal motion*, and is considered in more detail later. Here we may note that the phenomenon of heat is due to the thermal motion of molecules. The phenomenon in liquids or gases called "Brownian motion" shows that thermal motion is random.

If a cold spoon is dipped into hot tea, the collisions of molecules in contact transfer energy from the tea to the spoon. We call this energy *heat*. Heat continues to flow from the tea to the spoon until *thermal equilibrium* is reached. Conversely, if a hot spoon is dipped into cold tea, heat will flow from the spoon to the tea until thermal equilibrium is reached.

Which way heat will flow when two bodies A and B are placed in contact, depends on their respective *temperatures*. The temperature of a body is a measure of its hotness or coldness. Heat (energy) will flow from A to B when A is at a higher temperature than B. In this chapter we consider scales of temperature and the measurement of temperature.

Temperature scales. Absolute or thermodynamic scale

Any property of a substance which changes with temperature can be used to establish a *scale of temperature*. The mercury thermometer scale utilizes the change in volume of mercury with temperature. The platinum resistance thermometer scale utilizes the variation of electrical resistance of pure platinum with temperature (p. 7). The constant-volume gas thermometer scale utilizes the change of pressure with temperature of a fixed mass of gas at constant volume (p. 6).

All these temperature scales depend on the properties of a particular substance. The temperature scale chosen as the standard, however, is independent of the nature of the substance used. It is based on the second law of thermodynamics (p. 205) and is hence known as the *thermodynamic* (or *absolute*) *temperature scale*. On this scale temperatures are denoted by the symbol K, after Lord Kelvin who first suggested its use.

Any temperature scale must have two *fixed points*; here the temperature is always the same and easily reproducible. In the thermodynamic temperature scale, the *triple point of water* is one

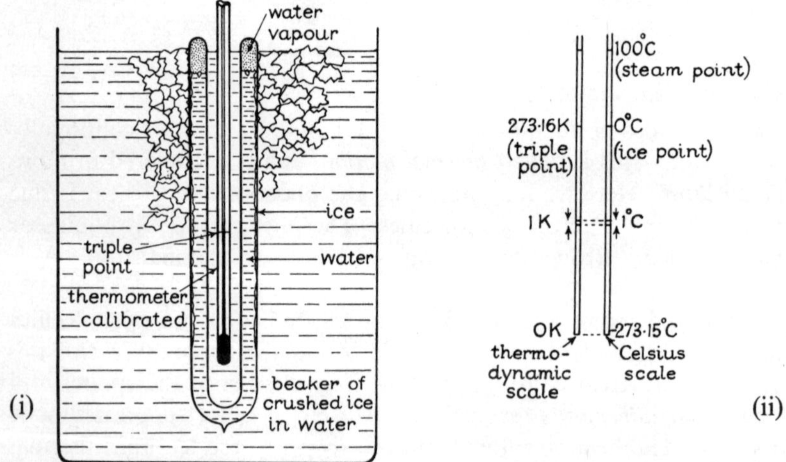

Fig. 1.1(i) Triple point measurement; (ii) Thermodynamic and Celsius scales

fixed point, whose temperature may be denoted by T_{tr}. It is the temperature at which ice, water and water-vapour coexist in equilibrium (p. 143), and this is *defined* as 273·16 K. Fig. 1.1 (i) illustrates a form of vessel suitable for the determination of the triple point of thermometers. The other or lower fixed point is the absolute zero, the temperature at which thermal motion vanishes. This is 0 K. The unit of thermodynamic temperature, the kelvin, is thus 1/273·16 *of the thermodynamic temperature of the triple point of water* (fig. 1.1 (ii)).

Celsius scale

On the Celsius (centigrade) scale of temperature, the lower fixed point is the temperature of melting ice at one atmosphere pressure and is defined as 0 °C. The upper fixed point is the temperature of steam above water boiling at one atmosphere pressure and defined as 100 °C. Measurements show that

$$0\ ^\circ\mathrm{C} = 273{\cdot}15\ \mathrm{K}, \quad 100\ ^\circ\mathrm{C} = 373{\cdot}15\ \mathrm{K}$$

Hence an interval or change of 1 °C *is equal to* 1 K (fig. 1.1 (ii)).

We shall use the symbol t to denote temperature on the Celsius scale and T to denote temperature on the thermodynamic scale. From above, it follows that t °C is changed to T K by the relation

$$T = 273{\cdot}15 + t$$

Thus 27 °C = 300·15 K or 300 K approximately.

Constant-volume gas thermometer scale

The *gas thermometer*, described on p. 6, utilizes the change in pressure with temperature of a fixed mass of gas at constant volume to provide a "gas thermometer temperature scale".

Suppose p_{T}, the pressure of the gas at an unknown temperature T, is found to be 780 mmHg, and p_{tr}, the pressure at the triple point, is 730 mmHg. Then, on the gas thermometer scale, T is given by

$$T = \frac{780}{730} \times 273{\cdot}16\ \mathrm{K} = 291{\cdot}9\ \mathrm{K}$$

Generally, a temperature T on the gas thermometer scale is calculated from the relation

$$T = \frac{p_{\mathrm{T}}}{p_{\mathrm{tr}}} \times 273{\cdot}16\ \mathrm{K}$$

If the ice point is used in elementary work, then T may be found from $T = (p_{\mathrm{T}}/p_{\mathrm{ice}}) \times 273{\cdot}15\ \mathrm{K}$.

Temperature on the Celsius scale

Suppose a platinum resistance thermometer is calibrated at only the ice and steam points. Then, since there are 100 °C between these two points, any other temperature t on the Celsius scale is given by

$$\frac{t}{100} = \frac{R_{\mathrm{t}} - R_{\mathrm{ice}}}{R_{\mathrm{st}} - R_{\mathrm{ice}}} \quad . \quad . \quad . \quad . \quad . \quad . \quad \text{(i)}$$

where R_t is the resistance at t °C, R_{st} is the resistance at the steam point and R_{ice} is the resistance at the ice point. The temperature T on the thermodynamic scale may then be found from $T = 273{\cdot}15 + t$.

Thus suppose the resistance R_t is measured at an unknown temperature t °C. If

$$R_{ice} = 6{\cdot}284 \text{ ohms}, R_{st} = 7{\cdot}321 \text{ ohms and } R_t = 6{\cdot}726 \text{ ohms},$$

then
$$\frac{t}{100} = \frac{6{\cdot}726 - 6{\cdot}284}{7{\cdot}321 - 6{\cdot}284},$$

from which $t = 42{\cdot}6$°C. Hence $T = 273{\cdot}15 + 42{\cdot}6 = 315{\cdot}8$ K.

If a constant-volume gas thermometer is calibrated only at the ice and steam points, a similar expression to (i) may be used in measuring an unknown temperature t °C. Thus

$$\frac{t}{100} = \frac{p_t - p_{ice}}{p_{st} - p_{ice}},$$

where p_t is the pressure at t °C, p_{st} is the pressure at the steam point and p_{ice} is the pressure at the ice point.

Example

The following readings were taken with a simple constant-volume air thermometer. This has a fixed mass of air trapped by a mercury column in a closed vertical limb, with the other vertical limb containing mercury and open to the atmosphere.

	Level of mercury in closed limb, mm	Level of mercury in open limb, mm
Bulb in melting ice	136	112
Bulb in steam at 760 mm pressure	136	390
Bulb at room temperature . .	136	160

Calculate the room temperature.

Suppose H is the barometric height in mmHg. Then, if T is the room temperature in K,

$$p_{ice} = H - (136 - 112) = H - 24 \quad \text{mmHg} \quad . \quad . \quad . \quad . \quad \text{(i)}$$
$$p_{st} = H + (390 - 136) = H + 254 \quad \text{mmHg} \quad . \quad . \quad . \quad . \quad \text{(ii)}$$
$$p_T = H + (160 - 136) = H + 24 \quad \text{mmHg} \quad . \quad . \quad . \quad . \quad \text{(iii)}$$

Now $\quad p_{st} - p_{ice} = 100 \text{ K}$

and $\quad p_T - p_{ice} = T - 273$

Hence, from (i), (ii), (iii),

$$\frac{T - 273}{100} = \frac{48}{278}$$

$$\therefore T = 290 \text{ K (approx)} = 17 \text{ °C}$$

Standard thermometer scale

At low pressures, all gases obey the same laws whatever their nature (p. 142). Consequently the constant-volume gas thermometer scale is chosen as the "standard" scale for measuring temperature. Temperatures measured on other thermometer scales are converted for comparison to temperatures on the gas thermometer scale.

If the temperature of a liquid is measured by a mercury thermometer, then by a constant-volume gas thermometer, and then by a platinum resistance thermometer, the results obtained are not the same; for example, the temperatures on the respective thermometer scales might be 28·2 °C, 28·0 °C, 28·1 °C. A little thought will show that this is not a surprising result. In one case the property used to measure temperature is the change in volume of mercury with temperature change; in another case the pressure variation with temperature of a gas at constant volume is used; and in the third case the variation of electrical resistance with temperature. *Thus the numerical value given to a temperature depends on the type of thermometer used.* The differences in the values are comparatively small in practice, and may be obtained from tables which have been compiled, using the temperature on the gas thermometer scale as a reference standard.

International temperature scale

In practice the standard thermometer, the gas thermometer, is often inconvenient for measuring temperature. A number of secondary fixed points have therefore been accurately measured by a gas thermometer and are listed below. This *international temperature scale* enables other thermometers to be calibrated and corrections applied.

(i)	Boiling-point of oxygen	−182·97 °C	(90·18 K)
(ii)	Ice point	0·000 °C	(273·15 K)
(iii)	Steam point	100·000 °C	(373·15 K)
(iv)	Boiling-point of sulphur	444·60 °C	(717·75 K)
(v)	Melting-point of antimony	630·5 °C	(903·6 K)
(vi)	Melting-point of silver	960·8 °C	(1233·9 K)
(vii)	Melting-point of gold	1063·0 °C	(1336·1 K)

The platinum resistance thermometer is used to measure temperatures between −190 °C and the melting-point of antimony, 630 °C. From −190 °C to 0 °C, the temperature is calculated from the formula

$R_t = R_0[1 + At + Bt^2 + C(t - 100)^3]$, where A, B are constants as below, and C is a constant determined from the oxygen point. From 0°C to 630°C, the temperarure is calculated from the formula $R_t = R_0[1 + At + Bt^2]$, where A, B are constants found from measurements at the ice point, the steam point and sulphur point. In the range 630°C to 1063°C, the temperature is found from the formula $E = a + bt + ct^2$, where E is the e.m.f. of a thermocouple made of pure platinum and platinum-rhodium metals and a, b, c are constants. Above 1063 °C, the temperature is found by optical pyrometry.

Constant-volume gas thermometer

The essential features of an accurate constant-volume gas thermometer are shown in fig. 1.2; it was designed by Harker and Chappuis. The bulb B may contain helium or nitrogen, and is connected by a capillary tube to a manometer M. The volume of the gas is kept constant in all measurements of pressure by bringing the mercury level to the tip of the pointer A_1. This is done by raising or lowering the tube T. A barometer NH is incorporated into the instrument so that the pressure of the gas can be directly determined at the place where the temperature is actually measured. Thus a vacuum exists above the mercury level in N.

When a measurement is taken, the level of the mercury in N is brought to the tip of a pointer A_2 by raising or lowering N. The reading on the scale S, which is at the same height as the tip of A_2,

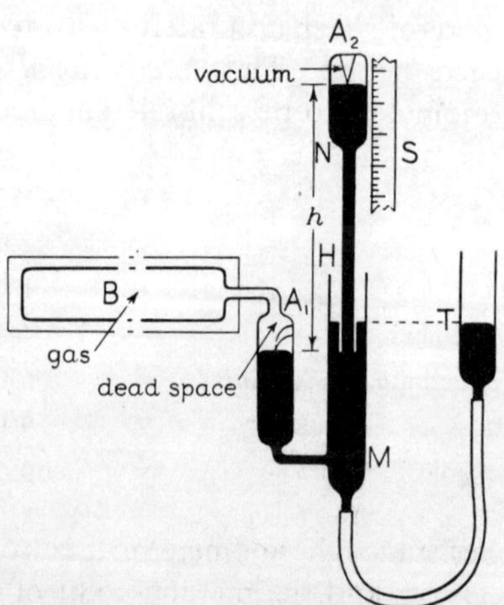

Fig. 1.2 Constant-volume gas thermometer

then corresponds to the level of mercury at the top of the barometer. The reading h on S corresponding to the tip of A_1 now gives a measure of the gas pressure in the bulb B. Only a very small percentage of the gas is in the space outside the bulb B, called the "dead space", so that only a very small correction is required for this gas which cannot be brought to the same temperature as that in B.

The laws obeyed by different gases differ slightly. At infinitely low pressures, however, all gases obey the same laws (see p. 142). In hydrogen and helium gas thermometers the gases are at low pressures, and extrapolation is made to zero pressure in "corrections" to these thermometers. Hydrogen is used for low temperatures to about −250 °C and helium to about −260 °C. Nitrogen is used at high temperatures to 1500 °C; hydrogen diffuses through containers at about 500 °C and cannot be used, therefore, for high-temperature measurements.

Platinum resistance thermometer

The electrical resistance of pure platinum increases regularly with increasing temperature. The resistance of a metal can be measured to a high degree of accuracy by means of a special Wheatstone bridge arrangement, and the *platinum resistance thermometer* is an accurate instrument for measuring temperature.

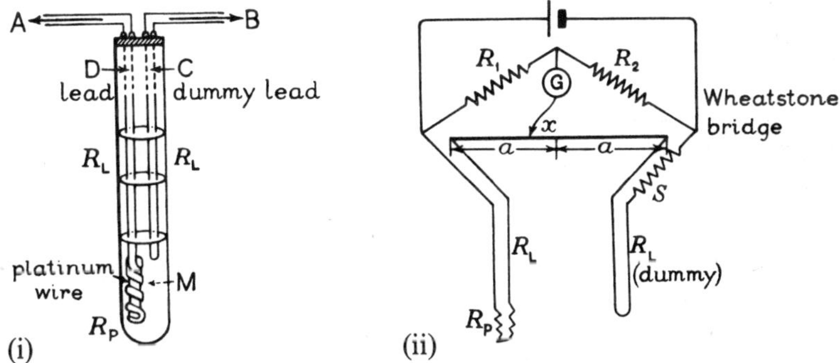

Fig. 1.3 Platinum resistance thermometer and bridge measurement

Fig. 1.3 (i) illustrates the essential features of a platinum resistance thermometer. A fine platinum wire is wound on a mica frame M, and is connected to one side A of a Wheatstone bridge arrangement by means of the leads D (fig. 1.3 (ii)). To neutralize the resistance introduced by D, "compensating" or "dummy" leads C are used which have the same resistance as D. The leads C are placed beside D, and are connected to the side B of the Wheatstone bridge *opposite* to A. The whole arrangement is placed inside a tube from which nearly all the air has been removed, and the leads are kept apart by discs of mica. To measure the resistance R_P of the platinum wire the leads are connected to a slide-wire Wheatstone bridge, shown in fig. 1.3 (ii). Then, at a balance,

$$\frac{R_1}{R_2} = \frac{R_P + R_L + (a - x)\rho}{R_L + S + (a + x)\rho}$$

where R_L is the total resistance of the leads, $2a$ is the length of the wire, x is the length of wire from the balance-point to the middle, ρ is the resistance per unit length of the wire, and S is a variable known resistance. In practice R_1 is made equal to R_2, and hence

$$R_P + R_L + (a - x)\rho = R_L + S + (a + x)\rho$$

$$\therefore R_P = S + 2x\rho$$

Thus the resistance of the leads is eliminated by the use of the "dummy" leads, and so the resistance measured is always that of the platinum wire alone.

The platinum resistance thermometer has the wide range of about −200 °C to 1200 °C. One disadvantage is the time required to make the bridge balance when measuring a temperature.

Thermoelectric thermometer

If two unlike metals A, B, such as copper and iron, are joined together, and one of the junctions H is heated, an e.m.f. is developed in one direction, say AHB, and a current flows (fig. 1.4 (i)). The thermoelectric e.m.f., as it is called, depends on the difference in

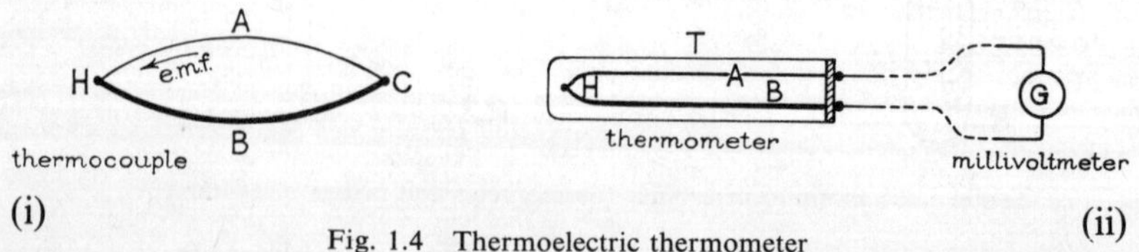

Fig. 1.4 Thermoelectric thermometer

temperatures between C, the "cold" junction, and H, the "hot" junction, and is obtained by a conversion of heat energy to electrical energy.

The *thermoelectric thermometer* utilizes the thermoelectric e.m.f. Fig. 1.4 (ii) represents a nickel wire A and a Nichrome wire B inside an iron tube T, with a high-resistance millivoltmeter G connected to A and B by very long leads. When the end of T is placed in contact with a hot object, an e.m.f. is developed in the circuit, and the temperature is read directly from the scale on G. This is calibrated previously in degrees with the aid of a platinum resistance or gas thermometer.

Nickel and Nichrome can be used for temperatures up to about 1200 °C. For higher temperatures the tube must be made of silica, and metals with high melting-points, such as tungsten and molybdenum, or platinum and a platinum-rhodium alloy, are used.

The junction of a thermocouple has a very low heat capacity. Hence it quickly reaches the temperature measured. The thermometer is thus useful for measuring varying temperatures. The small size of the junction also enables the temperature difference between parts of the same surface to be measured.

The thermoelectric thermometer has a wide range of −250 °C to 1600 °C. It is used in industry for measuring the temperatures of blast and glass furnaces.

Very low temperatures

When a substance is magnetized, energy is expended on it in orientating groups of particles or electron orbits. Conversely, on demagnetization energy is lost, and under adiabatic conditions, therefore, the temperature of the substance falls. F. Simon of Oxford and others have produced very low temperatures by using a paramagnetic salt such as potassium chrome alum or iron chrome alum, whose susceptibility χ varies considerably at these extreme temperatures. The salt D is suspended in a vessel surrounded by a Dewar flask containing liquid helium A, which is in turn surrounded by a Dewar flask containing liquid hydrogen B (fig. 1.5). Helium gas is first introduced round the salt, and after a time the temperature of the salt reaches about 1 K, the temperature of the surrounding helium. The whole apparatus is placed between the poles NS of a powerful electromagnet producing a field of the order of 1 T (10^4 gauss), and the current is switched on. The salt becomes warmed, and after a time the heat is conducted away by the surrounding helium gas, so that the temperature is again about 1 K. The helium gas is now pumped away, the current is switched off, and the salt becomes demagnetized under adiabatic conditions. The salt is thus cooled. Temperatures down to 0·0014 K have been reached by this method.

Measurements of such low temperatures are based on a magnetic law due to Curie which has been verified for a wide range of temperatures. This states that $\chi = C/T$, where χ is the susceptibility, T is the absolute temperature and C is a constant. At very low temperatures, therefore, χ is high. The susceptibility is measured either by a ballistic method or by an a.c. method using a mutual inductance

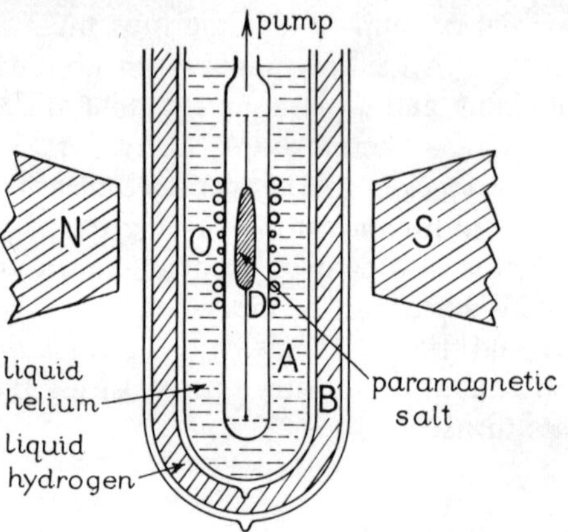

Fig. 1.5 Low-temperature production

bridge. In either case the sample is surrounded by two coils O, a primary of a few turns and a secondary of a large number of turns, as shown in fig. 1.5. The ballistic method is simpler, but individual readings cannot be taken more frequently than about once every 15 seconds. The apparatus in the a.c. method is more complicated, but it can give almost continuous readings of the susceptibility. This measurement of temperature, it will be noted, provides a "magnetic" scale of temperature, based on Curie's law. It is corrected to the Kelvin scale.

EXERCISE 1

1. The pressure in a constant-volume gas thermometer at the triple point of water, 273·16 K, is 720 mmHg. Calculate the temperatures on the thermodynamic temperature scale when the pressures are respectively (i) 770 mmHg, (ii) 40 mmHg.

2. A platinum resistance thermometer has a resistance of 2·50 ohms at the triple point of water. What is the temperature on the thermodynamic scale when the resistance is (i) 2·45 ohms, (ii) 2·85 ohms?

3. A thermocouple thermometer has an e.m.f. of 2·0 millivolts at the lower fixed point of the Celsius scale and an e.m.f. of 2·2 millivolts at the upper fixed point. Calculate the temperature on the Celsius scale when the e.m.f. is (i) 1·8 millivolts, (ii) 2·3 millivolts.

4. Explain how a Celsius scale of temperature may be defined in terms of a property of a substance. Discuss the desirable features of a thermometric property, illustrating your answer by reference to (*a*) the expansion of alcohol in a glass tube, (*b*) the resistance of platinum.

Describe how you would calibrate a constant-volume air thermometer and compare its reading with that of a mercury-in-glass thermometer at about 50 °C. (*L.*)

5. A bath of oil is maintained at a steady temperature of about 180 °C which is measured both with a platinum resistance thermometer and a mercury-in-glass thermometer. Explain why you would expect the temperatures indicated by the two thermometers to be different. At what temperatures would the two thermometers show the same value? (*N.*)

6. Explain what is meant by a *scale of temperature*. What type of thermometer would you use to measure the temperature of (i) a constant-temperature bath with dimensions of about 5 cm × 5 cm × 5 cm, at a temperature of about 50 °C, and (ii) the gases in a factory chimney which might vary over the range 250 °C–300 °C in a period of about a minute? Explain briefly how you would calibrate the instruments you describe. State the scale of temperature on which the result is obtained, and how you would convert it to the temperature on the ideal gas scale. (*O. and C.*)

7. Distinguish between *heat* and *temperature*, and explain what is meant by the statement that the temperature of a body is t °C on the scale of a certain thermometer.

Explain how you would use a piece of resistance wire as a thermometer, and describe how you would determine the readings corresponding to the fixed points of its scale.

A satellite orbits the earth outside its atmosphere. What factors determine the temperature of the outer shell of the satellite (*a*) on the side exposed to the sun, (*b*) on the side remote from the sun? (*C.*)

8. Explain the principle underlying the establishment of a centigrade temperature scale in terms of some suitable physical property.

What type of thermometer would you choose for use in experiments involving (*a*) the plotting of a cooling curve for naphthalene in the region of its melting-point, (*b*) finding the boiling-point of oxygen, (*c*) the measurement of the thermal conductivity of a small crystal? In each instance give reasons for your choice.

If the resistance R_t of the element of a resistance thermometer at a temperature of t °C on the ideal gas scale is given by $R_t = R_0(1 + At + Bt^2)$, where R_0 is the resistance at 0 °C and A and B are constants such that $A = -6{\cdot}50 \times 10^3 B$, what will be the temperature on the scale of the resistance thermometer when $t = 50{\cdot}0$ °C? (*L.*)

9. How is a scale of temperature defined? What is meant by a temperature of 15 °C?

On what evidence do you accept the statement that there is an absolute zero of temperature at about −273 °C?

In a special type of thermometer a fixed mass of gas has a volume of 100·0 cm^3 and a pressure of 81·6 cm of mercury at the ice point, and volume 124·0 units with pressure 90·0 units at the steam point. What is the temperature when its volume is 120·0 units and pressure 85·0 units, and what value does the scale of this thermometer give for absolute zero? Explain the principle of your calculation. (*O. and C.*)

10. Give a brief account of the principles underlying the establishment of a scale of temperature and explain precisely what is meant by the statements that the temperature of a certain body is (*a*) t °C on the constant-volume air scale, (*b*) t_p °C on the platinum resistance scale, and (*c*) t_T °C on the Cu-Fe thermocouple scale. Why are these three temperatures usually different?

Describe an optical pyrometer and explain how it is used to measure the temperature of a furnace. (*N*.) [See also p. 193.]

11. Explain what is meant by a change in temperature of 1 °C on the scale of a platinum resistance thermometer.

Draw and label a diagram of a platinum resistance thermometer together with a circuit in which it is used.

Give *two* advantages of this thermometer and explain why, in its normal form, it is unsuited for measurement of varying temperatures.

The resistance R_t of platinum varies with the temperature t °C as measured by a constant-volume gas thermometer according to the equation

$$R_t = R_0(1 + 8000\alpha t - \alpha t^2)$$

where α is a constant. Calculate the temperature on the platinum scale corresponding to 400 °C on this gas scale. (*N*.)

2

Heat Energy. Heat Capacity. Latent Heat

Heat is a form of energy

The exact nature of heat was a problem which occupied the minds of many early scientists. Until the beginning of the nineteenth century heat was considered to be a material substance. Thus a body gaining heat was thought to increase its content of "caloric", as the material substance was called, and an object losing heat lost some "caloric".

In 1798, however, Benjamin Thompson, or Count Rumford as he became known, recorded an observation which could not be explained by the caloric theory. At that time he was engaged in superintending the boring of a gun barrel at a munitions factory in Bavaria. He noticed that a large amount of heat was generated, although the quantity of metal chips per minute obtained from the gun barrel was very small indeed. He thought it extremely unlikely that such a small quantity of metal could have had stored in it such a large quantity of heat (caloric). He also observed that the longer the boring took place, the greater was the amount of heat produced. Rumford thus associated the heat produced with the *mechanical energy* expended by the borer. To verify this theory, Rumford bored a gun barrel for some hours in a tank of water, and to the astonishment of observers, who could see that no flame was used, the water was eventually made to boil.

Rumford communicated his observations to the Royal Society, but little notice was taken of them. In 1842, however, J. P. Joule of Manchester began a series of famous experiments which proved conclusively that heat is not a material substance. He converted different forms of energy, such as mechanical and electrical energy, into heat, using specially designed apparatus. Each time he measured the quantity of heat produced and the amount of energy expended and found they were always in a constant ratio. He therefore concluded that *heat is a form of energy*. For example, the molecules of a solid vibrate with greater amplitude about their mean positions when the solid is heated, so that their total energy is increased (p. 54). Also, the molecules of a liquid and a gas move faster when heated, and hence the kinetic energy of their molecules is increased (p. 59).

Quantity of heat. Joule and watt

Since heat is a form of energy, it was decided in 1948 to measure quantity of heat in *joules*, symbol J. The joule is the SI unit of heat.

The calorie (cal) was originally the unit of heat; it is the amount of heat required to raise the temperature of one gramme of water from 14·5 to 15·5 °C. Experiment shows that 1 cal = 4·1868 J, or, approximately, 4·2 J = 1 cal. A larger unit of heat widely used is the *kilojoule*, symbol kJ, which is 1000 J. One kilocalorie (kcal) = 1000 cal = 4200 J (approx) = 4·2 kJ.

The *rate* at which heat is gained or lost by a body is expressed in joules per second or *watts*, W. A hot metal block which loses 600 J in 30 s has an average rate of loss of heat of 20 J s^{-1} or 20 W.

Heat (thermal) capacity. Specific heat capacity

The heat required to raise the temperature of a body by 1 °C is known as the *heat* (*thermal*) *capacity* of the body. Heat capacity, symbol C, is thus expressed in "joules per K", symbol J K^{-1}, since 1 °C = 1 K.

The heat required to raise the temperature of unit mass of a substance by 1 °C is known as the *specific heat capacity* of the substance, symbol c. Specific heat capacity may be expressed in joules per kg per K symbol "J kg^{-1} K^{-1}". The specific heat capacity of water is about 4·2 kJ kg^{-1} K^{-1} (4·2 J g^{-1} K^{-1}), and this will be denoted by c_W. The specific heat capacity of copper is about 0·4 kJ kg^{-1} K^{-1} or 400 J kg^{-1} K^{-1}.

If the mass of a substance is m, its specific heat capacity is c and its rise in temperature t, then, from the definition of specific heat capacity, the quantity of heat Q supplied is given by

$$Q = mct$$

Thus if a quantity of heat Q of 10 kJ is supplied to 2 kg of copper of $c = 0{\cdot}4\ \mathrm{kJ\ kg^{-1}\ K^{-1}}$, the temperature rise t of the copper is given by

$$t = \frac{Q}{mc} = \frac{10}{2 \times 0{\cdot}4} = 12{\cdot}5\ \mathrm{K}$$

Note that, with a unit of mass of 1 g instead of 1 kg, $c = 0{\cdot}4\ \mathrm{J\ g^{-1}\ K^{-1}}$. In this case, since $Q = 10\ \mathrm{kJ} = 10\,000\ \mathrm{J}$, and mass of copper $= 2\ \mathrm{kg} = 2000\ \mathrm{g}$, then

$$t = \frac{Q}{mc} = \frac{10\,000}{2000 \times 0{\cdot}4} = 12{\cdot}5\ \mathrm{K}$$

the same result as before.

Electrical method for specific heat capacity

Electrical methods of measuring specific heat capacity are straightforward and capable of providing accurate results.

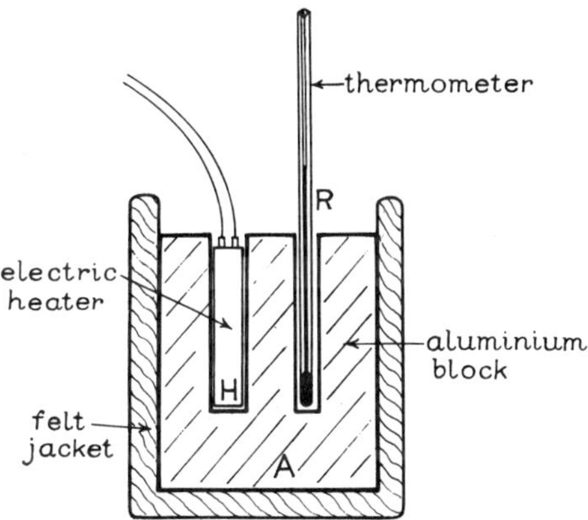

Fig. 2.1 Laboratory electrical method for specific heat capacity

Fig. 2.1 illustrates the basic principle of measuring the specific heat capacity of a *metal* A. A polished aluminium block, mass 1 kg, is placed in an insulating felt jacket, and an electrical heater H and a thermometer R are placed inside holes drilled in the block. After a known time, the final maximum temperature rise is observed.

The following results were obtained:

Heater supply = 50 W. Time = 5 min. Temperature rise = 14 °C.

$$\therefore \text{heat supplied} = 50 \times (5 \times 60)\ \mathrm{J} = 15\,000\ \mathrm{J}$$

$$\therefore \text{specific heat capacity } c = \frac{15\,000\ \mathrm{J}}{1\ \mathrm{kg} \times 14\ °\mathrm{C}} = 1100\ \mathrm{J\ kg^{-1}\ K^{-1}}\ (\text{approx})$$

$$= 1{\cdot}1\ \mathrm{kJ\ kg^{-1}\ K^{-1}}$$

Specific heat capacity of solids by electrical method

About 1911 Nernst and Lindemann designed a vacuum calorimeter for measuring accurately the specific heat capacity of metals, especially at low temperatures (fig. 2.2). A cylindrical block of the metal A was drilled, and a cylindrical plug B of the same metal was fitted in. B was shaped at the top so that good thermal contact with A was made. A platinum wire, insulated with paraffin wax, was wound round B, and not only enabled the metal to be electrically heated, but also provided, by its resistance measurement, the initial and final temperatures of the block.

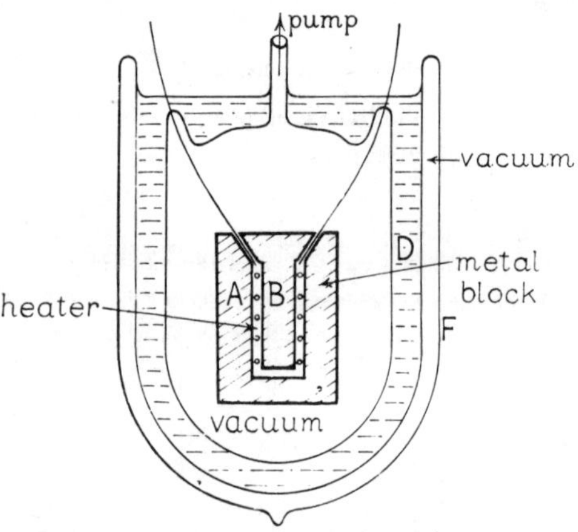

Fig. 2.2 Specific heat capacity by electrical method

The whole arrangement was suspended in a glass flask D, inside a Dewar flask filled with ice or solid carbon dioxide or liquid air or liquid hydrogen, according to the low temperature required. Hydrogen was first admitted into D, and the metal eventually assumed the same temperature as the surroundings. The hydrogen was then pumped off, so that no heat was now lost by conduction or convection; the current was switched on, and the resistance of the coil was measured when conditions were steady. A small rise in temperature, about 1 °C, was arranged. In this way Nernst found the variation of the specific heat capacity of a metal for temperatures ranging from −256 °C to 0 °C.

Specific heat capacity of liquid by electrical method

The specific heat capacity of a liquid such as water can be measured by an electrical method, using a so-called "continuous-flow" method.

The essential features of the apparatus are shown in fig. 2.3. The liquid whose specific heat capacity is required flows steadily through a glass tube AB from A to B. During its passage it is heated electrically by a coil of wire placed along the length of the tube. After a time the inlet and outlet temperatures of the water become constant, and *under these steady conditions*, the difference in temperature is measured by means of platinum resistance thermometers P, Q.

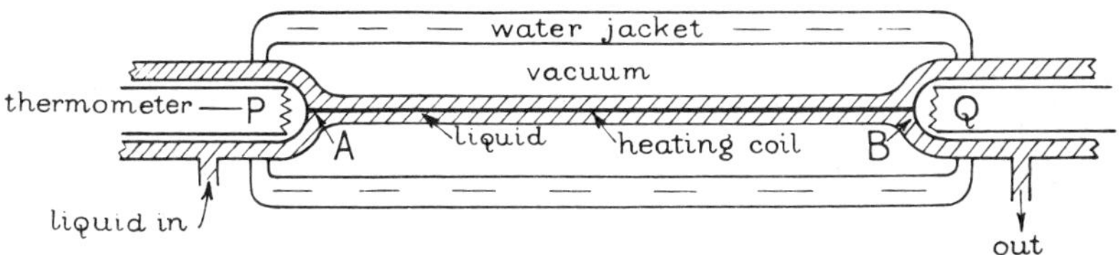

Fig. 2.3 Continuous-flow calorimeter for liquid specific heat capacity

Suppose I is the current in amperes in this wire, V is the potential difference in volts between A and B, m is the mass of liquid flowing steadily along the tube per second, and t is the steady difference in temperature of the water at A and B. When conditions are steady, no heat is expended on the glass flow tube or calorimeter since its temperature is constant. *All* the heat produced by the electric current is then used in heating the liquid as it flows from A to B. Now the electrical energy expended per second is IV. If h is the small amount of heat lost per second by radiation through the vacuum, then

$$IV = mct + h \quad . \quad . \quad . \quad . \quad . \quad . \quad \text{(i)}$$

where c is the specific heat capacity of the liquid.

To eliminate h, the steady flow of liquid is now decreased to a new value. At the same time the current is also diminished until the *same* liquid temperatures are obtained at A and B. The heat lost by radiation is then the same (the surrounding water-jacket (fig. 2.3) is maintained at the same temperature), and if I_1, V_1, m_1 denote the new magnitudes of current, p.d., and mass of liquid per second, we have

$$I_1 V_1 = m_1 ct + h \quad . \quad . \quad . \quad . \quad . \quad . \quad \text{(ii)}$$

Subtracting (ii) from (i),

$$\therefore IV - I_1 V_1 = (m - m_1)ct$$

$$\therefore c = \frac{IV - I_1 V_1}{(m - m_1)t}$$

Advantages

The advantages of the method and of the "continuous-flow" calorimeter, as it is called, are:

(i) the heat capacity of the calorimeter, a doubtful quantity, does not need to be considered;

(ii) the difference in temperature between A and B can be measured very accurately because conditions are steady;

(iii) the cooling correction is negligible except for very accurate work;

(iv) V and I may be measured accurately by a potentiometer method;

(v) *the method can be used for investigating the variation of the specific heat capacity of a liquid with temperature.* Here the difference in temperature is made small, e.g. the temperature at A may be 32 °C and that at B 34 °C, in which case the specific heat capacity c is determined at the average temperature 33 °C. The electrical method has been used to determine the variation of the specific heat capacity of water with temperature.

Specific heat capacity by mechanical method

Mechanical energy can be supplied to a solid copper cylinder by winding a few turns of nylon cord round it and attaching weights of about 5 kg to one end of the cord (fig. 2.4). The other end of the cord

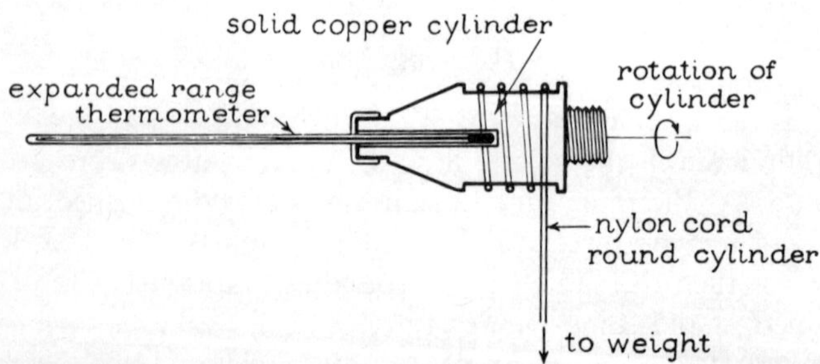

Fig. 2.4 Mechanical energy method for specific heat capacity

is attached to a fixed elastic band, which, with adjustment, goes slack when the cylinder is rotated steadily, so that the weights are completely supported by the cord. The tension in the cord is then equal to the total weight.

Suppose a steady rotation of 330 revolutions produces a final maximum temperature rise of 16 °C when the tension in the chord is the weight of 5 kg or $5 \times 9{\cdot}8 = 49$ newtons. If the diameter of the solid copper cylinder is 2·54 cm, then, since 1 newton × 1 metre = 1 joule,

$$\text{mechanical work done } W = \text{tension} \times \text{circumference} \times \text{no. of revs}$$
$$= 49 \times (\pi \times 2{\cdot}54 \times 330/100)\text{ J}$$
$$= 1{\cdot}3 \times 10^3\text{ J (approx)}$$

If the cylinder has a mass of 185 g or 0·185 kg,

$$\text{specific heat capacity} = \frac{W}{\text{mass} \times \text{temp. rise}} = \frac{1{\cdot}3 \times 10^3}{0{\cdot}185 \times 16}\text{ J kg}^{-1}\text{ K}^{-1}$$
$$= 440\text{ J kg}^{-1}\text{ K}^{-1}\text{ (approx)} = 0{\cdot}44\text{ kJ kg}^{-1}\text{ K}^{-1}$$

Example

An electrical heating coil is incorporated in the copper cylinder of fig. 2.4, and a current of 0·3 A produces a temperature rise of 2 °C in 15 min. Calculate the p.d. across the coil.

From the mechanical work calculation, a temperature rise of 16 °C in the copper cylinder is due to a supply of energy of $1{\cdot}3 \times 10^3$ J.

$$\therefore 2\,°\text{C temperature rise is due to } \frac{2}{16} \times 1{\cdot}3 \times 10^3\text{ J}$$

If V is the p.d. in volts across the coil, electrical energy supplied

$$= IVt = 0{\cdot}3 \times V \times (15 \times 60)\text{ J}$$
$$\therefore 0{\cdot}3 \times V \times 900 = \frac{2}{16} \times 1{\cdot}3 \times 10^3$$
$$\therefore V = 0{\cdot}6\text{ V (approx)}$$

Method of mixtures

If a solid is relatively small, its heat capacity, and the specific heat capacity of its material, can be found by a method of "mixtures". In this case the solid is heated in boiling water or a steam chamber, and then quickly transferred to cold water inside a metal calorimeter. The final water temperature is then observed.

Suppose m, c are the respective mass and specific heat capacity of the solid; m_W, c_W the corresponding quantities for water and m_1, c_1 for the calorimeter; t is the initial temperature of the hot solid and t_1, t_2 are the initial and final temperatures of the water. Assuming no heat is lost, then

$$\text{heat lost by solid} = \text{heat gained by water and calorimeter}$$
$$\therefore mc(t - t_2) = m_W c_W(t_2 - t_1) + m_1 c_1(t_2 - t_1)$$

Thus knowing $c_W(4{\cdot}2\ \mathrm{kJ\ kg^{-1}\ K^{-1}})$ and c_1 from tables, the specific heat capacity c of the solid can be calculated.

Provided no chemical action occurs, the method of mixtures can be used to find the specific heat capacity c_2 of a liquid. In this case the specific heat capacity c of the solid is known from tables, and the liquid is used in place of water inside the calorimeter.

Example

200 g of copper, specific heat capacity $0{\cdot}4\ \mathrm{kJ\,kg^{-1}\,K^{-1}}$, is heated to 100 °C. It is then quickly transferred to 80 g of water at 10 °C inside a calorimeter of heat capacity $100\ \mathrm{J\,K^{-1}}$. Calculate the final water temperature ($c_w = 4{\cdot}2\ \mathrm{kJ\,kg^{-1}\,K^{-1}}$).

Let x °C = final water temperature

Heat lost by copper $= 0{\cdot}2 \times 400 \times (100 - x)$ J.

Heat gained by water and calorimeter $= (0{\cdot}08 \times 4200 \times \overline{x - 10}) + (100 \times \overline{x - 10})$ J

$$\therefore\ 0{\cdot}2 \times 400 \times (100 - x) = 0{\cdot}08 \times 4200 \times (x - 10) + 100(x - 10).$$

Solving for x, $\therefore\ x = 24$ °C

Cooling correction

When a hot solid is dropped into a liquid inside a calorimeter, as in the method of mixtures, the temperature of the liquid begins to rise above that of the surroundings. The liquid and calorimeter then lose heat to the surroundings. It therefore follows that the *observed* final temperature is less than the final temperature had no heat been lost. The temperature to be added to the observed final temperature to compensate for loss of heat is known as the "cooling correction". A cooling correction is necessary for accuracy in the method of mixtures.

One method of avoiding a cooling correction is to cool the liquid in the calorimeter below the surroundings before the hot solid is dropped into it. Suppose a trial run shows that a temperature rise of about 8 °C occurs. Then the liquid is cooled about 4 °C below the surroundings and the hot solid raises its temperature to about the same amount above. On the average, then, the temperature of the liquid and calorimeter has been about the same as that of the surroundings during the experiment. Very little net heat is then lost to the surroundings.

Newton's law of cooling

As the result of experiments, Newton stated that *the rate of loss of heat of a body by cooling is proportional to its excess temperature over the temperature of the surroundings*. This is known as *Newton's law of cooling*. Later experiments have shown that this law is an approximate one when the cooling body is surrounded by still air, but it holds at all temperatures when a draught is set up near the cooling body, i.e. under conditions of "forced convection" of the air. In the case of still air, Newton's law is fairly true for excess temperatures over the surroundings of not more than about 30°, a condition usually obtained in the method of mixtures, as the final temperature of the mixture is frequently less than 30° above the temperature of the surroundings.

Newton's law of cooling is expressed mathematically by $dQ/dt = -k\theta$, where θ is the temperature excess over the surroundings and k is a constant. Thus if m is the mass of the cooling body and c its specific heat capacity,

$$mc\frac{d\theta}{dt} = -k\theta$$

$$\therefore \int_{\theta_0}^{\theta} \frac{d\theta}{\theta} = -\frac{k}{mc}\int_0^t dt$$

where θ_0 is the excess temperature over the surroundings at $t = 0$.

$$\therefore \log_e\left(\frac{\theta}{\theta_0}\right) = -\frac{k}{mc}t$$

from which

$$\theta = \theta_0 e^{-(kt/mc)} \qquad \text{(i)}$$

Thus the temperature excess over the surroundings diminishes exponentially with time for a given body.

In still air, when *natural convection* takes place, Dulong and Petit found that the rate of loss of heat of a body in a constant-temperature enclosure was proportional to $\theta^{5/4}$, where θ is the excess temperature over the surroundings. This is sometimes known as the "five-fourths power law of cooling". It can be shown by plotting the logarithm of the heat lost per second by a given body against $\log \theta$, when a fair straight line of slope about 1·25 is obtained.

Dulong and Petit's law has been verified experimentally for temperature excesses of several hundred degrees down to about one degree. Lorentz has deduced the law mathematically.

Cooling-correction method

Figure 2.5 illustrates one method of applying a cooling correction. As the hot solid is dropped into the calorimeter, the temperature of the liquid is observed at equal intervals of time. The temperature

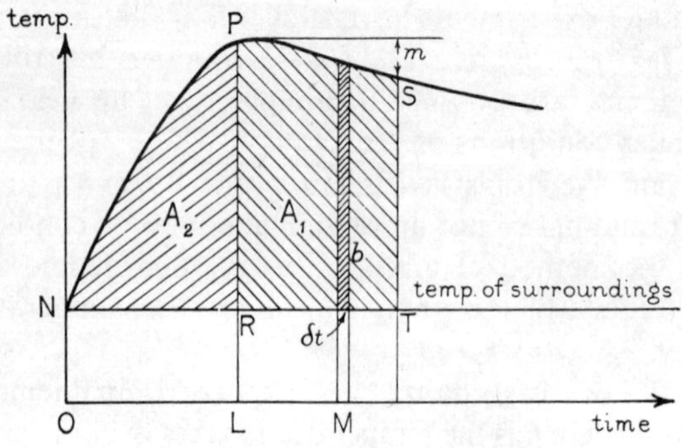

Fig. 2.5 Cooling correction

then rises with time along the graph NP to a maximum temperature at P, after which it falls slowly along PS. If m is the difference in temperature between the points corresponding to P and S, it is shown below that the cooling correction d is given by $d/A_2 = m/A_1$, where A_2, A_1 are the respective areas of NPR and RPST. Thus $d = mA_2/A_1$. By plotting the temperature-time graph on graph paper and counting squares, the ratio A_2/A_1 can be found and d calculated. The maximum temperature of the liquid, corrected for cooling, is then (*temperature at* $P + d$).

The proof of the formula $d/A_2 = m/A_1$ depends on Newton's law of cooling. Consider a temperature excess b over the surroundings, corresponding to a time M. From Newton's law the heat lost per second is kb, where k is a constant, and hence the heat lost in a further small time δt is $kb \,.\, \delta t$. But $b \,.\, \delta t =$ the small area δA shown shaded in fig. 2.5. Thus the heat lost $= k \,.\, \delta A$. If the total thermal capacity of the water and calorimeter is denoted by C, and $\delta\theta$ is the small decrease in temperature in the time δt, the heat lost $= C \,.\, \delta\theta$. Hence $C \,.\, \delta\theta = k \,.\, \delta A$. Integrating both sides, from the time corresponding to P to that corresponding to S, we obtain

$$Cm = kA_1 \quad \text{(i)}$$

For the period of time between O and L, it can be similarly shown that $C \,.\, d = k \,.\, A_2$, as d is the temperature change due to *cooling* in this time. Thus, with (i),

$$\frac{Cd}{Cm} = \frac{kA_2}{kA_1} \quad \text{or} \quad \frac{d}{m} = \frac{A_2}{A_1}. \quad \text{Hence } d = \frac{mA_2}{A_1}$$

A rough value for a cooling correction can be obtained by observing the time OL or NR taken to reach the observed maximum temperature P, and then noting the temperature drop from P in *half* this time. In this case RT in fig. 2.5 $= \frac{1}{2}$NR. Now $A_2 =$ area of triangle NPR (approximately) $= \frac{1}{2} \,.\,$ NR . RP; and $A_1 =$ area of trapezium RTSP $= \frac{1}{2}$(RP + ST) . RT $= \frac{1}{2} \times$ (RP + RP) . RT, to a rough approximation $= \frac{1}{2} \times$ RP $\times$ 2RT. But 2RT = NR. Hence

$A_2 = A_1$, and thus $d = mA_2/A_1 = m$. Consequently, *the cooling correction d is roughly the temperature drop from the observed maximum in half the time taken to reach the maximum temperature.*

Specific heat capacity of a liquid by method of cooling

When a hot body is placed inside an enclosure at a lower constant temperature, the heat lost per second by the hot body depends on

(i) *the excess temperature over the enclosure at the instant considered,* and
(ii) *the nature and area of the exposed surface of the body.*

It should be noted by the reader that this statement is true whatever the temperature and the hot body may be, and it should be carefully distinguished from Newton's law of cooling (p. 20).

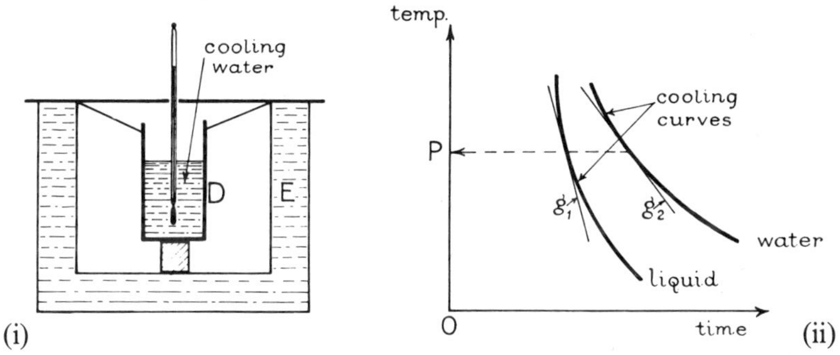

Fig. 2.6 Specific heat capacity by method of cooling

Suppose that a known volume of hot liquid is placed inside a blackened calorimeter D of known heat capacity C, which is suspended in a constant-temperature enclosure E (fig. 2.6 (i)). To prevent loss of heat by evaporation a lid is placed on the top, and the temperature of the liquid is noted at equal intervals of time while it cools. A cooling curve is then plotted. The experiment is now repeated with the same calorimeter containing an *equal volume* of hot water, and a cooling curve is plotted (fig. 2.6 (ii)). Now at the same temperature, the heat lost per minute by the calorimeter and water is equal to the heat lost per minute by the calorimeter and liquid, since the nature and area of the cooling bodies are the same in each case, and so are the respective temperatures. Thus

$$(m_2 c_W + C)g_2 = (m_1 c + C)g_1 \qquad \text{(i)}$$

where m_2, m_1 are the masses of the water and liquid respectively, c_W, c are their respective specific heat capacities, C is the heat capacity of the calorimeter, and g_2, g_1 are the respective losses in temperature per minute of the water and liquid at the temperature P under consideration. The values of g_2, g_1 are found from the *slopes of the tangents* to the cooling curves corresponding to the temperature P, as shown in fig. 2.6. Since m_2, C, m_1, g_1, g_2 are known, the unknown specific heat capacity c of the liquid can be calculated from (i).

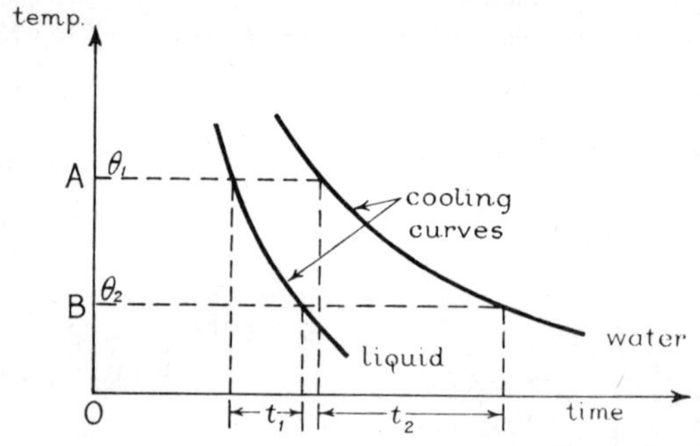

Fig. 2.7 Cooling curves

Another method can be used to calculate the specific heat c of the liquid from the two cooling curves. In this method the respective times t_1, t_2 to cool from a temperature θ_1(A) to a temperature θ_2(B) are noted from the liquid and water cooling curves (see fig. 2.7). Since the temperatures θ_1, θ_2 are the same in both cases, and the same calorimeter and volumes of water and liquid are used, the average heat lost per minute is the same. Thus

$$\frac{(m_2 c_W + C)(\theta_1 - \theta_2)}{t_2} = \frac{(m_1 c + C)(\theta_1 - \theta_2)}{t_1}$$

$$\therefore \frac{m_2 c_W + C}{t_2} = \frac{m_1 c + C}{t_1}$$

Since m_2, C, m_1, t_2, t_1 and c_W are all known, then c can be calculated.

The reader should carefully note that the *volumes* of the liquid and water used in the experiment are the same; thus leading to similar conditions in the two cases.

Examples

1. State Newton's law of cooling. Distinguish clearly between the "rate of cooling" and the "rate of loss of heat" of a vessel containing a liquid at a temperature higher than that of its surroundings.

200 g of water, and an equal volume of another liquid of mass 250 g, are placed in turn in the same calorimeter of mass 100 g and specific heat capacity 0·4 kJ kg^{-1} K^{-1} The liquids, which are constantly stirred, are found to cool from 60 °C to 40 °C in 180 s and in 140 s respectively. Find the specific heat capacity of the liquid. (*W.*)

Second part.—The average heat lost per second by the liquid and calorimeter while the temperature falls from 60 °C to 40 °C is equal to the average heat lost per second by the water and calorimeter from 60 °C to 40 °C, since the excess temperature over the surroundings, and the nature and area of the cooling surfaces, are the same in both cases. As pointed out on p. 23, it should be noted that we are not using Newton's law of cooling.

The average rate of loss of heat of the liquid and calorimeter in joules per second $= (0{\cdot}25c + 0{\cdot}1 \times 400)(60 - 40)/140$, and that of the water and calorimeter $=$ $(0{\cdot}2 \times 4200 + 0{\cdot}1 \times 400)(60 - 40)/180$.

$$\therefore \frac{(0{\cdot}25c + 40)20}{140} = \frac{880 \times 20}{180}$$

Solving, $\therefore c = 2{\cdot}6 \text{ kJ kg}^{-1} \text{ K}^{-1}$ (approx)

2. State Newton's law of cooling, and explain how you would test it experimentally.

A copper calorimeter of mass 100 g, containing 150 cm^3 of liquid of specific heat capacity 2·5 kJ kg^{-1} K^{-1} and relative density 1·2, is found to cool at the rate of 2 °C per minute when its temperature is 50 °C above that of its surroundings. If the liquid is emptied out and 150 cm^3 of a liquid of specific heat capacity 1·7 kJ kg^{-1} K^{-1} and relative density 0·9 are substituted, what will be the rate of cooling when the temperature is 40 °C above that of the surroundings? (Specific heat capacity of copper = 0·4 kJ kg^{-1} K^{-1}.) (*O. and C.*)

Since density of water = 1·0 g cm^{-3}, density of liquid = 1·2 g cm^{-3}. Hence mass = 150 × 1·2 g = 180 g = 0·18 kg. Also, specific heat capacity = 2500 J kg^{-1} K^{-1}.

For calorimeter, mass = 100 g = 0·1 kg. Also, specific heat capacity = 400 J kg^{-1} K^{-1}.

∴ total heat lost per minute at 50 °C above surroundings =

$$(0{\cdot}18 \times 2500 + 0{\cdot}1 \times 400)2 \text{ J} \quad \ldots \quad (1)$$

For second liquid, mass = 150 × 0·9 = 135 g = 0·135 kg; specific heat capacity = 1700 J kg^{-1} K^{-1}. If x °C per minute is the rate of cooling at 40 °C above surroundings, then

$$\text{total heat lost per minute} = (0{\cdot}135 \times 1700 + 0{\cdot}1 \times 400)x \text{ J} \quad \ldots \quad (2)$$

Assuming Newton's law of cooling,

$$\frac{(0{\cdot}135 \times 1700 + 0{\cdot}1 \times 400)x}{(0{\cdot}18 \times 2500 + 0{\cdot}1 \times 400)2} = \frac{40}{50}$$

$$\therefore \frac{269{\cdot}5x}{980} = \frac{40}{50}$$

$$\therefore x = 2{\cdot}9 \text{ °C min}^{-1} \text{ (approx)}$$

Heat energy.
Heat capacity.
26 Latent heat

LATENT HEAT

Fusion

If some ice in a calorimeter is gently warmed and stirred, and the temperature is noted throughout, the thermometer will remain steady at 0 °C until *all* the ice has melted to water. As further heat is supplied the thermometer indicates an increased reading, so that the temperature of the water formed now rises.

While the temperature was constant at 0 °C, the heat supplied was melting, or *fusing*, the ice. Usually, the application of heat to a body results in a rise of its temperature; but there is no such effect when a solid melts, and the heat supplied was given the name of "hidden" or *latent* heat years ago when the phenomenon was not understood. The heat required to change a solid at its melting-point to liquid at the melting-point is called the *latent heat of fusion*.

We have already seen that heat is a form of energy (p. 13). The latent heat of fusion is the energy needed to overcome the intermolecular forces or bonds between all the molecules which maintain the solid state. The molecules are then free to move through the volume of liquid formed.

Specific latent heat of fusion

The *specific latent heat of fusion* of a substance is defined as *the quantity of heat required to change unit mass from a solid to a liquid at the melting-point*. The specific latent heat will be denoted by the symbol l; the unit is $\mathrm{J\,kg^{-1}}$ with a unit mass of 1 kg. The specific latent heat of fusion of ice is about $334\,\mathrm{kJ\,kg^{-1}}$ (or $334\,\mathrm{J\,g^{-1}}$). Thus 334 kJ of heat is given out when 1 kg of water at 0°C changes to ice at 0°C, and 334 kJ of heat is absorbed when 1 kg of ice at 0°C is melted at 0 °C to form water at 0 °C. The specific latent heat of fusion of aluminium is about $450\,\mathrm{kJ\,kg^{-1}}$ and its melting-point about 660 °C.

Suppose m kg of ice at 0 °C is changed to water at 0 °C. Then ml kJ of heat is needed, where l is in $\mathrm{kJ\,kg^{-1}}$. If the ice is changed to water at 15 °C, an additional amount of heat is needed to warm the water from 0 °C to 15 °C. This amount is $m \times 4{\cdot}2 \times (15 - 0)$ kJ, since the specific heat capacity of water c_W is $4{\cdot}2\,\mathrm{kJ\,kg^{-1}\,K^{-1}}$, or $63m$ kJ. Hence the *total* heat supplied to m kg of ice to change it to water at 15°C, using $l = 334\,\mathrm{kJ\,kg^{-1}}$,

$$= (ml + 63m)\,\mathrm{kJ} = 397m\,\mathrm{kJ}$$

Determination of specific latent heat of fusion of ice

The specific latent heat of fusion l of ice can be simply and accurately found from the method of mixtures. Pieces of ice are dried and dropped into a calorimeter containing water, and the final temperature is noted after all the ice has melted. The cooling correction can be eliminated by using warm water whose temperature is as much above the room temperature as the final temperature is below (p. 21).

As an illustration of the calculation for l, suppose the following measurements were taken:

Mass of copper calorimeter = 47·3 g, specific heat capacity of copper = 400 J kg^{-1} K^{-1}

Mass of water = 75·7 g, specific heat capacity of water = 4200 J kg^{-1} K^{-1}

Mass of ice = 11·0 g

Initial water temperature = 18·4 °C; final temperature = 6·4 °C. Then

$$\text{heat lost by calorimeter} = (47{\cdot}3 \times 10^{-3}) \times 400 \times (18{\cdot}4 - 6{\cdot}4)\ \text{J}$$

$$\text{heat lost by water} = (75{\cdot}7 \times 10^{-3}) \times 4200 \times (18{\cdot}4 - 6{\cdot}4)\ \text{J}$$

heat gained by ice in changing to water at 6·4 °C

$$= 11 \times 10^{-3} \times l + 11 \times 10^{-3} \times 4200 \times (6{\cdot}4 - 0)\ \text{J}$$

where l is in J kg^{-1}. Assuming no heat losses,

$$\therefore\ 11 \times 10^{-3}\, l + 295{\cdot}7 = (47{\cdot}3 \times 0{\cdot}4 + 75{\cdot}7 \times 4{\cdot}2)(18{\cdot}4 - 6{\cdot}4)$$

Solving,

$$\therefore\ l = 340 \times 10^{3}\ \text{J kg}^{-1} = 340\ \text{kJ kg}^{-1}$$

Determination of specific heat capacity of liquid using ice

We have already mentioned that the specific heat capacity of a liquid can be found by a method of mixtures (p. 19). A method which is particularly useful for inflammable liquids is one in which ice is dropped into the liquid, and the final temperature is observed after melting. Thus, if m is the mass of liquid, c its specific heat capacity, C the heat capacity of the calorimeter, t_1 and t the initial and final temperatures of the liquid, then the heat lost $= (mc + C)(t_1 - t)$. If the mass of ice used is m', the heat required to change the ice to water at t°C $= m'l + m'c_{\text{W}}(t - 0)$, where l is the specific latent heat of fusion of ice. Thus

$$(mc + C)(t_1 - t) = m'l + m'c_{\text{W}}t$$

As l and other quantities are known, c can be calculated.

Bunsen's ice calorimeter

Bunsen invented an "ice calorimeter" for measuring specific heat capacities. This consists essentially of a tube B containing air-free water to a level C; mercury fills the lower portion, which leads to a horizontal capillary tube D, fig. 2.8. A tube A is fixed inside B, and the whole arrangement (except D) is placed inside a vessel containing ice. Some ether (not shown) is placed inside A, and air is bubbled through the ether, which then evaporates. The heat absorbed from the surrounding water results in a coating of ice all round the end of A, and the apparatus is then placed inside a double-walled container of melting ice (fig. 2.8), so that the space all round A and B is at a temperature of 0 °C.

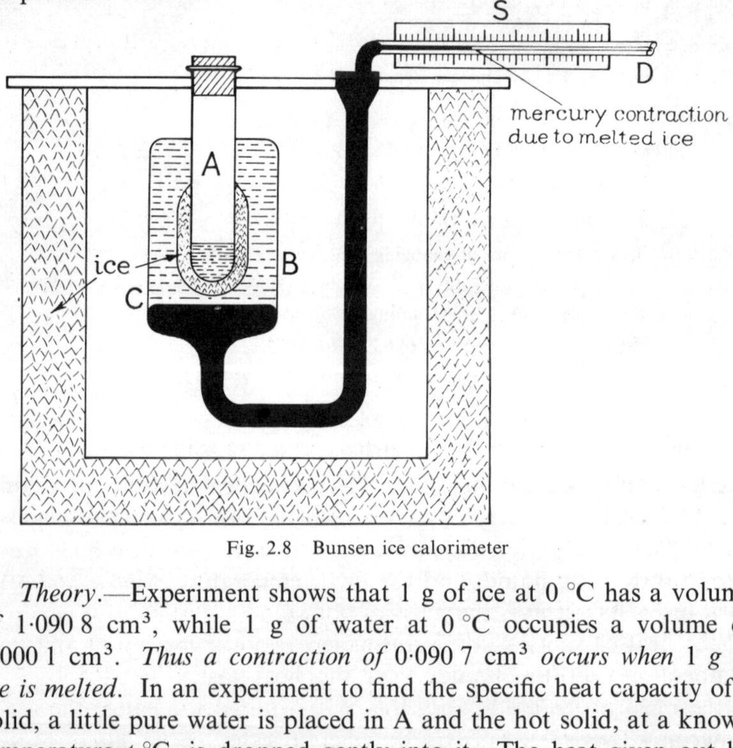

Fig. 2.8 Bunsen ice calorimeter

Theory.—Experiment shows that 1 g of ice at 0 °C has a volume of 1·090 8 cm^3, while 1 g of water at 0 °C occupies a volume of 1·000 1 cm^3. *Thus a contraction of* 0·090 7 cm^3 *occurs when* 1 *g of ice is melted.* In an experiment to find the specific heat capacity of a solid, a little pure water is placed in A and the hot solid, at a known temperature t °C, is dropped gently into it. The heat given out by the solid when its temperature falls to 0 °C causes a little ice to melt, and the mercury column in D contracts by an amount which is measured by the scale S.

Suppose v is the contraction in volume in cm^3; m and c are the mass and specific heat capacity of the solid; l is the specific latent heat of fusion of ice. Since a contraction of 0·090 7 cm^3 indicates that 1 g of ice has melted, it follows that $v/0{\cdot}090\ 7$ g of ice, or $10^{-3}\ v/0{\cdot}090\ 7$ kg, has been melted. The heat required for this purpose is $10^{-3}\ vl/0{\cdot}090\ 7$

joules, and this is the heat given out by the solid when its temperature falls from t °C to 0 °C. It follows that

$$mct = \frac{10^{-3}vl}{0{\cdot}090\,7}$$

$$\therefore\ c = \frac{10^{-3}vl}{0{\cdot}090\,7mt}$$

Thus knowing m, t, v, l, the specific heat capacity can be calculated.

This method of measuring specific heat capacity depends on a knowledge of l, the specific latent heat of fusion of ice. Another method can be adopted which does not depend on a knowledge of l. In this case a known mass M of pure water at a known temperature t_1°C is added to the water originally in A. The mercury column then recedes by a number of divisions, n say, which thus corresponds to a quantity of heat Mc_Wt_1 joules. In this way the number of joules per division is found. The heat Q given up by a hot solid when placed in A can now be found easily, and its specific heat capacity is obtained from $c = Q/(mt)$ if m is the mass and t is the original temperature.

As the mercury column recedes by an appreciable amount for small quantities of heat, Bunsen, and later experimenters in 1947, were able to measure the specific heats of rare metals, which were available only in small quantities. An advantage of his calorimeter is that no cooling correction is required, as the surroundings are at 0 °C, the same temperature as the calorimeter A. On the other hand, the ice round A may be below the assumed temperature 0 °C owing to the pressure exerted on it by the surrounding water and mercury, because the melting-point of ice decreases with pressure. This is a serious disadvantage of the ice calorimeter for very accurate work.

Melting-point. Freezing-point

When a pure solid is heated, it usually melts at a definite temperature. The temperature remains constant until all the solid is melted, after which the liquid temperature rises as heat is further supplied (see p. 26). If the hot liquid is now allowed to cool and the temperature observed at equal time intervals, the cooling curve ABCD is obtained (fig. 2.9 (i)). The flat part BC corresponds to the *freezing-point* (F.P.) of the solid. At B, the latent heat of fusion given out per second when the liquid freezes compensates exactly for the heat lost per second to the surroundings. The temperature is then constant, as shown.

The melting-point of a substance can thus be deduced from its

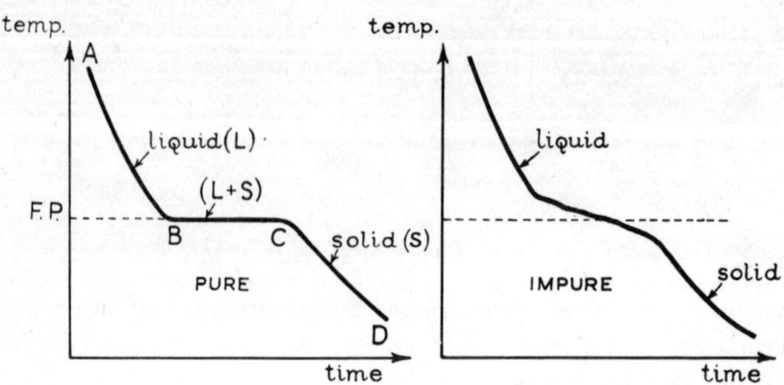

Fig. 2.9 Melting-point from cooling curve

temperature-time curve as it cools from the liquid to the solid state; it is the temperature which corresponds to the flat portion of the curve. In the case of substances with high melting-points, such as metals, the mercury thermometer is unsuitable, and a thermocouple or pyrometer (p. 191) must be used.

There is no definite melting-point in the case of *amorphous* or *plastic* substances, such as glass, paraffin wax, and sealing wax, as solidification takes place over a range of temperature. The cooling curve is then roughly as illustrated in fig. 2.9 (ii). The melting-point of an *alloy* varies with its composition, and is usually lower than the melting-point of any of its constituents.

Effect of pressure on melting-point

The change from the liquid to the solid state is usually accompanied by a change in volume of the substance. Most substances diminish in volume when they solidify. Water, however, is an exception to the rule as it expands when it changes to ice, 1 cm^3 of water forming about 1·09 cm^3 of ice (p. 28). The great force exerted in expansion when water freezes leads to burst pipes in winter.

The melting-point of a substance is affected by changes in pressure. We can deduce what will happen from elementary considerations. Take the case of ice, for example, which melts at 0 °C under normal atmospheric pressure (760 mm of mercury). If the pressure on the ice increases, the conditions are more favourable to the formation of water because ice contracts when it melts, and so 0 °C is above the new melting-point. For one atmosphere increase in pressure the melting-point of ice is lowered by about 0·0075 °C. Increased pressure *lowers* the melting-point of all substances which *contract* when they change from the solid to the liquid state. For substances which expand when they change from the solid to the liquid state, increased pressure raises the melting-point.

The formation of a snowball when the snow is compressed is explained by the fact that increased pressure lowers the melting-point of ice. Some of the snow melts because its temperature is then above the new melting-point, and when the pressure is diminished, the water formed freezes to ice again and binds the particles of snow together. Tyndall made an impressive demonstration of *regelation* or re-freezing. A thin wire was wound round a block of ice and a heavy weight was attached to the wire below the ice. The high pressure of the wire on the top of the ice made the ice melt, and as the wire sank into the ice, the water formed re-froze above the wire, where the pressure was reduced. In this way the wire slowly sank through the solid block of ice without any apparent "cutting" of the ice, and finally emerged on the other side.

Vaporization. Latent heat

When water is heated, its temperature rises until it begins to boil. At this stage the water changes from the liquid to the vapour state. Steam is then obtained, and the temperature of the water remains constant at about 100 °C. The heat now supplied is called the "latent heat of vaporization". It is the energy needed to overcome completely the bonds of attraction between the molecules in the liquid state and the external pressure (p. 55). The molecules then escape from the liquid and exist in the gaseous state, where they move about fairly independently of each other in the space occupied.

The *specific latent heat of vaporization* (l) of a substance is defined as the quantity of heat required to change unit mass from a liquid at the boiling-point to vapour at the same temperature. The unit of l is J kg^{-1}, or, with large values of l, kJ kg^{-1}.

The specific latent heat of vaporization of water is about 2260 kJ kg^{-1} ($2260\,\text{J g}^{-1}$). Thus if m kg of steam at 100°C is condensed to water at 100°C, the heat given out $= ml = m \times 2260 \times 10^3$ J or $m \times 2260$ kJ. If m kg at 100°C is condensed to water at 30°C, an additional quantity of heat is given out. This is equal to $mc_W(100-30)$ joules, where c_W is the specific heat capacity of the water, $= m \times 4200 \times (100 - 30)\ \text{J} = m \times 4{\cdot}2 \times (100 - 30)$ kJ. Thus the total amount of heat given out

$$= (m \times 2260 + m \times 294)\ \text{kJ} = 2554m\ \text{kJ}$$

Determination of specific latent heat of steam

The specific latent heat of steam can be obtained by the method of mixtures. Some water is boiled in a flask, and using a "trap" for drops of water, steam alone is obtained issuing from the trap. The steam is passed into water placed inside a calorimeter, and the final temperature of water is observed. The mass of steam condensed is obtained by re-weighing the calorimeter and its contents. To illustrate the calculation, suppose the following measurements are taken:

Mass of calorimeter = 47·5 g, specific heat capacity = $400\ \mathrm{J\,kg^{-1}\,K^{-1}}$

Mass of water = 122·6 g, specific heat capacity = $4200\ \mathrm{J\,kg^{-1}\,K^{-1}}$

Mass of steam = 4·1 g, temperature = 100 °C

Initial water temperature = 15·0 °C, final temperature = 33·0 °C

Heat gained by water and calorimeter

$$= (122{\cdot}6 \times 10^{-3} \times 4200 + 47{\cdot}5 \times 10^{-3} \times 400)(33{\cdot}0 - 15{\cdot}0)\ \mathrm{J}$$

$$= (122{\cdot}6 \times 4{\cdot}2 + 47{\cdot}5 \times 0{\cdot}4)18\ \mathrm{J}$$

Heat given up by steam in condensing to water at 33·0 °C

$$= 4{\cdot}1 \times 10^{-3}\, l + 4{\cdot}1 \times 10^{-3} \times 4200 \times (100 - 33)\ \mathrm{J}$$

where l is in $\mathrm{J\,kg^{-1}}$.

$$\therefore 4{\cdot}1 \times 10^{-3}\, l + 4{\cdot}1 \times 4{\cdot}2 \times 67 = (122{\cdot}6 \times 4{\cdot}2 + 47{\cdot}5 \times 0{\cdot}4)\ 18$$

Solving, $\therefore l = 2060 \times 10^{3}\ \mathrm{J\,kg^{-1}} = 2060\ \mathrm{J\,kg^{-1}}$

Electrical method for specific latent heat of vaporization

One of the most accurate methods of measuring the latent heat of vaporization is by an electrical method whose *principle* is illustrated in fig. 2.10 (i). D is an air-tight vessel containing the liquid, which is electrically heated by a current passing through S into a heater coil A. T is a platinum resistance thermometer for measuring the temperature of the vapour obtained, which passes down a tube E into a vessel B where it may be collected. After a time, when conditions are steady, the vapour is obtained at a steady rate and the mass of liquid per second m collected in B is measured. Then the heat supplied per second is ml, where l is the specific latent heat of vaporization, as the liquid in D is at the boiling-point when vapour is obtained. But the heat supplied per second is IV, where I is the current in A, and V is the p.d. across A (p. 17).

$$\therefore IV = ml + h \quad \ldots\ldots \quad \text{(i)}$$

where h is the heat per second lost from the vessel at the boiling-point. To eliminate h, the experiment is repeated with a faster rate of

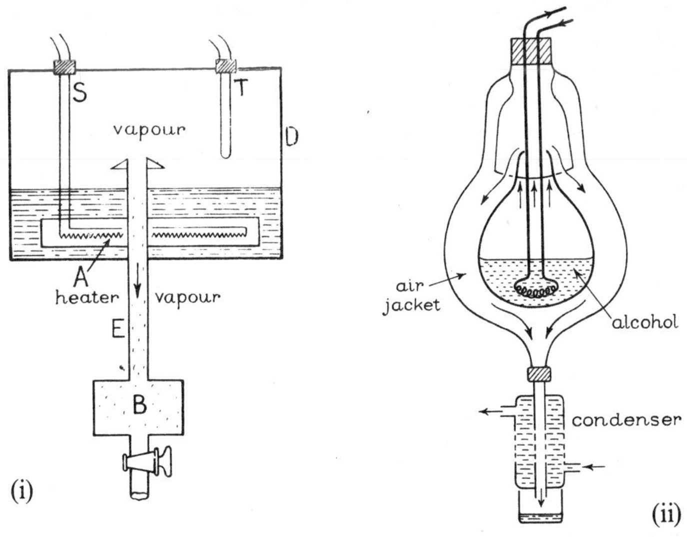

Fig. 2.10 Electrical method for l

heating, with an increased current I_1 and p.d. V_1. Suppose m_1 is the increased mass per second of liquid collected. Then, since h is the same,

$$I_1 V_1 = m_1 l + h \quad . \quad . \quad . \quad . \quad . \quad . \quad . \quad \text{(ii)}$$

Subtracting (i) from (ii),

$$I_1 V_1 - IV = (m_1 - m)l$$

$$\therefore l = \frac{I_1 V_1 - IV}{m_1 - m}$$

In this expression, l is in J kg^{-1} when I, I_1 are in amperes, V, V_1 in volts and m in kg s^{-1}.

The variation of l with temperature can be investigated by altering the pressure in D (see p. 127). Figure 2.10 (ii) illustrates a school apparatus for measuring the specific latent heat of a liquid such as alcohol by an electrical method.

Internal energy and latent heat

When 1 cm^3 of water at 100 °C is heated, about 1671 cm^3 of steam at 100 °C are formed. From p. 98, the energy required to overcome the external atmospheric pressure = pressure × volume change. Assuming the pressure p is standard pressure,

then $$p = 1{\cdot}013 \times 10^5\,\mathrm{N\,m^{-2}}$$

and

$$\text{volume change} = 1671 - 1 = 1670\,\mathrm{cm^3\,g^{-1}} = 1670 \times 10^3\,\mathrm{cm^3\,kg^{-1}}$$
$$= 1{\cdot}67\,\mathrm{m^3\,kg^{-1}}$$

$$\therefore \text{energy to overcome external pressure} = 1{\cdot}013 \times 10^5 \times 1{\cdot}67\,\mathrm{J\,kg^{-1}}$$
$$= 169\,\mathrm{kJ\,kg^{-1}} \quad \ldots \quad \text{(i)}$$

This is sometimes called the "external part" l_{ext} of the specific latent heat value. To find the "internal part" l_{int} of the specific latent heat, or the energy required to completely break the bonds between liquid molecules so that a gas is formed, we subtract l_{ext} from l, the measured specific latent heat of steam, which is about 2260 kJ kg^{-1} at standard pressure. Thus

$$l_{int} = 2260 - 169 = 2090\,\mathrm{kJ\,kg^{-1}}\ \text{(approx)}$$

Since $l_{ext} = 169\,\mathrm{kJ\,kg^{-1}}$, it follows that the energy required to break the bonds with neighbours is about 12 times as great as the energy needed to overcome the external atmospheric pressure.

Examples

1. Draw a labelled diagram to illustrate an experiment for determining the specific latent heat of vaporization of water. Do *not* describe the experiment, but state how the result would be calculated from the experimental data.

Water in a vacuum flask is boiled steadily by passing an electric current through a coil of wire immersed in the water. When the potential difference across the coil is 5·25 V and the current through it is 2·58 A, 6·85 g of water evaporate in 20 min. When the potential difference and the current are maintained at 3·20 V and 1·57 A respectively, 2·38 g of water evaporate in 20 min, all the other conditions being the same. Calculate the specific latent heat of vaporization of water. (*N.*)

If I is the current in amperes in a coil of wire and V is the potential difference in volts across the coil, the heat produced *per second* in the coil is IV joules (p. 32). This heat is expended in evaporating the water and in making up the heat losses. Thus, firstly,

$$2{\cdot}58 \times 5{\cdot}25 = \frac{6{\cdot}85 \times 10^{-3}\,l}{20 \times 60} + h \quad \ldots \quad \text{(i)}$$

where l is the specific latent heat of vaporization in J kg^{-1}, and h is the heat lost per second, and 20 min = 20 × 60 s. In the second case,

$$1{\cdot}57 \times 3{\cdot}20 = \frac{2{\cdot}38 \times 10^{-3}\,l}{20 \times 60} + h \quad \ldots \quad \text{(ii)}$$

as the conditions of cooling are the same. Subtracting (ii) from (i) to eliminate h, we have

$$(2{\cdot}58 \times 5{\cdot}25 - 1{\cdot}57 \times 3{\cdot}20) = \frac{10^{-3}(6{\cdot}85 - 2{\cdot}38)}{20 \times 60} l$$

$$\therefore l = \frac{1200(2{\cdot}58 \times 5{\cdot}25 - 1{\cdot}57 \times 3{\cdot}20)}{10^{-3} \times 4{\cdot}47}$$

$$= 2290 \times 10^3 \text{ J kg}^{-1} = 2290 \text{ kJ kg}^{-1}$$

2. Describe the Bunsen ice-calorimeter and discuss its merits.

The capillary tube of such a calorimeter has an internal diameter of 0·4 mm. When a piece of metal of mass 0·5 g heated to 100 °C is dropped into the calorimeter, the mercury meniscus moves 4 cm. What is the specific heat capacity of the metal? (Relative density of water at 0 °C = 1·000, that of ice at 0 °C = 0·917; specific latent heat of fusion of ice = 336 kJ kg^{-1}.) (*O. and C.*)

First part.—See text.

Second part.—1 g of water has a volume of 1·000 cm^3; 1 g of ice has a volume of 1/0·917 cm^3 or 1·091 cm^3. Thus a contraction of 0·091 cm^3 occurs when 1 g of ice melts. Since the contraction $= \pi r^2 l = \pi \times 0{\cdot}02^2 \times 4 = 0{\cdot}0016\,\pi$ cm^3, the mass of ice melted $= 0{\cdot}0016\pi/0{\cdot}091$ g.

$$\therefore \text{heat given out by metal} = \frac{0{\cdot}0016\pi \times 336 \times 10^{-3}}{0{\cdot}091} \text{ kJ}$$

$$\therefore \text{specific heat capacity of metal} = \frac{Q}{mt} = \frac{0{\cdot}0016\pi \times 336 \times 10^{-3}}{0{\cdot}091 \times 0{\cdot}5 \times 10^{-3} \times 100}$$

$$= 0{\cdot}37 \text{ kJ kg}^{-1} \text{ K}^{-1}$$

3. Describe and explain the method you would use to find the specific latent heat of evaporation of alcohol.

A copper calorimeter of mass 120 g contains 150 g of a liquid at 26 °C. 25 g of ice at 0 °C are added, and the minimum temperature of the mixture is recorded. At this temperature, vapour from the same boiling liquid is introduced into the mixture until the final temperature of the calorimeter and new content is 30 °C. What mass of vapour is condensed? Take the specific heat capacity of copper as 0·4 kJ kg^{-1} K^{-1}, the specific heat capacity of the liquid as 2·4 kJ kg^{-1} K^{-1}, the boiling-point of the liquid as 64·7 °C, the specific latent heat of evaporation of the liquid as 1100 kJ kg^{-1}, the specific latent heat of fusion of ice as 320 kJ kg^{-1}. (*L.*)

Let t °C = the minimum temperature of the mixture.

Since heat lost by liquid and calorimeter = heat gained by ice and water formed,

$$(150 \times 10^{-3} \times 2400 + 120 \times 10^{-3} \times 400)(26 - t)$$
$$= 25 \times 10^{-3} \times 320 \times 10^3 + 25 \times 10^{-3} \times 4200 \times (t - 0),$$

from which $t = 5{\cdot}1$ °C (approx)

Let m = mass in kg of vapour condensed. Since heat given out by condensed vapour and liquid formed = heat gained by liquid, water, and calorimeter,

$$\therefore 1100 \times 10^3 m + m \times 2400 \times (64{\cdot}7 - 30{\cdot}0)$$
$$= (150 \times 10^{-3} \times 2400 + 25 \times 10^{-3} \times 4200 + 120 \times 10^{-3} \times 400)(30 - 5{\cdot}1)$$

$$\therefore 1183 \times 10^3 m = 12\,774$$

$$\therefore m = 10{\cdot}8 \times 10^{-3} \text{ kg} = 10{\cdot}8 \text{ g}$$

EXERCISE 2

(Where necessary, assume $c_W = 4{\cdot}2\,\text{kJ}\,\text{kg}^{-1}\,\text{K}^{-1}$)

1. Calculate (i) the temperature rise of 200 g of copper, $c = 0{\cdot}4\,\text{kJ}\,\text{kg}^{-1}\,\text{K}^{-1}$, when 2·0 kJ of heat is given to it, (ii) the heat gained by 250 g of water when its temperature rises from 20 °C to 60 °C.

2. A calorimeter has a mass of 200 g and a specific heat capacity $0{\cdot}4\,\text{kJ}\,\text{kg}^{-1}\,\text{K}^{-1}$. What is its heat capacity? If the calorimeter contains 100 g of water at 10 °C, find the new temperature when 10 kJ is supplied to the water and calorimeter by a burner.

3. A lead bullet, mass 1·5 g, has a speed of 20 m s^{-1} and is brought to rest in a target. If 20% of the heat energy is gained by the bullet, calculate its temperature rise. (Specific heat capacity of lead = $0{\cdot}12\,\text{kJ}\,\text{kg}^{-1}\,\text{K}^{-1}$.)

4. An electric current of 2 A flows in a resistance wire whose p.d. is 50 V. Calculate the heat produced in 7 min. If the wire is fully immersed in 250 g of water initially at 20 °C, how long will the water take to reach boiling-point under normal atmospheric pressure?

5. A horizontal solid metal cylindrical calorimeter has a radius 2 cm. It is rotated about its axis for 2000 rev against a constant tension of 50 N in a rope wound round it. The temperature rise of the calorimeter is 8 °C and its mass is 400 g. If 20% of the heat produced is assumed to warm the calorimeter, estimate the specific heat capacity of its material.

6. (i) 250 g of water at 15 °C is heated until it all evaporates, the atmospheric pressure being 760 mmHg. Calculate the heat supplied. (ii) 3 g of steam of 100 °C is condensed and the final temperature of the water formed is 35 °C. Find the total heat given up. (l for steam = $2268\,\text{kJ}\,\text{kg}^{-1}$.)

7. (i) In melting to water at 8 °C, 5 g of ice at 0 °C requires 1800 J of heat. Calculate the specific latent heat of fusion of ice. (ii) What mass of iron at 17 °C, dropped into liquid oxygen at its boiling-point −183 °C, will cause 2 g of oxygen to evaporate? (l for oxygen = $210\,\text{kJ}\,\text{kg}^{-1}$, specific heat capacity of iron = $0{\cdot}4\,\text{kJ}\,\text{kg}^{-1}\,\text{K}^{-1}$.)

8. 5·8 g of steam at 100 °C is passed into a copper calorimeter containing 85 g of water at 16 °C. If the final temperature is 50 °C, calculate the mass of the calorimeter. (l for steam = $2268\,\text{kJ}\,\text{kg}^{-1}$; specific heat capacity of copper = $0{\cdot}4\,\text{kJ}\,\text{kg}^{-1}\,\text{K}^{-1}$.)

9. 100 cm^3 of alcohol takes 6 min to cool in a draught from 60 °C to 20 °C. An equal volume of water in the same calorimeter takes 18 min to cool from 60 °C to 20 °C in a draught. If the calorimeter heat capacity is $42\,\text{J}\,\text{K}^{-1}$, calculate a value for the specific heat capacity of alcohol, whose density is 0·8 g cm^{-3}.

10. Define *specific latent heat of vaporization*. Describe a method for determining the specific latent heat of vaporization of a liquid such as alcohol.

Steam, initially at 120 °C, is passed into 120 kg of solid grease, initially at 15 °C, until all the grease is just melted. The specific heat capacity of the grease is 2·50 kJ $\text{kg}^{-1}\,\text{K}^{-1}$, its specific latent heat of fusion is $175\,\text{kJ}\,\text{kg}^{-1}$ and its melting-point is 55 °C. Find the mass of steam required if the specific heat capacity of steam may be taken as $2{\cdot}00\,\text{kJ}\,\text{kg}^{-1}\,\text{K}^{-1}$, its specific latent heat is $2250\,\text{kJ}\,\text{kg}^{-1}$, and one-quarter of the heat imparted is lost from the grease. Assume the boiling-point of water under a pressure of one atmosphere to be 100 °C, and the specific heat capacity of water to be $4{\cdot}20\,\text{kJ}\,\text{kg}^{-1}\,\text{K}^{-1}$. (*L.*)

11. The unit of heat used to be the *calorie*; it is now the *joule*. Why has this change been made?

100 g of water are placed in a calorimeter fitted with an electrical heater. When the heater dissipates 10·6 watts, the temperature of the water rises to a steady value

below the prevailing boiling-point. With the heater switched off the rate of fall of temperature at this temperature is found to be 1·34 °C min^{-1}. The experiment is repeated with 100 g of another liquid in the same calorimeter. With a dissipation of 6·0 watts a steady temperature below the boiling-point of this liquid is attained, and at this temperature with the heater switched off the rate of fall of temperature is 1·00 °C min^{-1}. Calculate a value for the specific heat capacity of the liquid, assuming the specific heat capacity of water to be 4·20 kJ kg^{-1} K^{-1}. (*N.*)

12. Explain what is meant by the statement "4·2 joules = 1 calorie".

Give a labelled diagram of a continuous-flow apparatus which could be used to determine the specific heat of water. The diagram should include the electrical circuit, but a description of the apparatus is *not* required.

In such an experiment the following readings were taken:

Current in heating coil	2·0 A	1·5 A
Potential difference across coil	6·0 V	4·5 V
Mass of water collected	42·3 g	70·2 g
Time of flow	60 s	180 s
Inlet temperature	38·0 °C	38·0 °C
Outlet temperature	42·0 °C	42·0 °C

Explain how each reading would be taken, and use the figures to obtain a value for the specific heat capacity of water in J kg^{-1} K^{-1}. If each temperature reading was subject to an uncertainty of ±0·1 °C, find the resulting percentage uncertainty in the specific heat capacity due to this cause alone. (*C.*)

13. Explain why the temperature of a boiling liquid remains constant although heat energy is supplied to it. Describe briefly how you would measure the latent heat of vaporization of a liquid which has a boiling-point of about 80 °C and explain the precautions that must be taken to achieve an accurate result. (*O. and C.*)

14. Describe, with the aid of a labelled diagram, how you would find the specific heat capacity of a liquid by the method of continuous flow.

Discuss the advantages and disadvantages of the method compared with the method of mixtures.

The temperature of 50 g of a liquid contained in a calorimeter is raised from 15·0 °C (room temperature) to 45·0 °C in 530 seconds by an electric heater dissipating 10·0 watts. When 100 g of liquid is used and the same change in temperature occurs in the same time, the power of the heater is 16·1 watts. Calculate the specific heat capacity of the liquid. (*N.*)

15. State *Newton's law of cooling* and describe how to obtain observations and how to use them in order to test the validity of the law.

A solid of mass 250 g in a vessel of thermal capacity 67·2 J K^{-1} is heated to a few degrees above its melting-point and allowed to cool in steady conditions until solid again. Sketch the graph of its temperature plotted against time. If this graph shows that immediately before solidification starts the rate of cooling is 3·2 °C min^{-1}, while immediately after solidification it is 4·7 °C min^{-1}, calculate the specific heat capacity of the solid and the time taken by the solidifying process. (The specific latent heat of fusion of the substance may be taken as 146·6 kJ kg^{-1} and its specific heat capacity in the liquid state as 1·22 kJ kg^{-1} K^{-1}.) (*L.*)

16. Discuss the reasons for preferring electrical methods to the method of mixtures in finding an accurate value of a specific heat capacity.

A carbon block of mass 150 g was heated electrically in air. The equilibrium temperatures θ_e attained at each value of the heater power were as follows:

W (watts)	0	25	50	75	100	150	200	250
θ_e (°C)	25	190	240	285	325	390	440	480

The block was then allowed to cool, and the following temperature readings θ were obtained at intervals of time t:

t (min)	0	2	4	6	8	10	12	14	16	18
θ (°C)	500	395	315	265	230	200	180	160	145	130

Find the specific heat capacity of carbon at 150 °C, 250 °C, and 400 °C. (*C.*)

17. Distinguish between *specific heat capacity* and *specific latent heat*. With what physical changes is each associated? Describe the processes involved in terms of simple molecular theory.

A thin-walled tube containing 5 cm^3 of ether is surrounded by a jacket of water calibrated so that changes in the volume of the water can be read off. The whole apparatus is cooled down to 0 °C and all the ether is then evaporated by blowing a rapid stream of air pre-cooled to 0 °C through it. The change of volume as ice forms in the water is 0·35 cm^3. Calculate the specific latent heat of evaporation of the ether.

(Use the following values: specific latent heat of fusion of ice = 334 kJ kg^{-1}. Densities at 0 °C: water, 1·000 g cm^{-3}; ice, 0·917 g cm^{-3}; ether, 0·736 g cm^{-3}.) (*O. and C.*)

18. Discuss the nature of the heat energy (*a*) of a solid, (*b*) of a gas, (*c*) of the sun during the transmission of this energy to the earth.

265,430 joules of heat are produced when a vehicle of total mass 1270 kg is brought to rest on a level road. Calculate the speed of the vehicle in km per hr just before the brakes are applied. (*L.*)

19. The specific heat capacity of gallium metal is 0·33 kJ kg^{-1} K^{-1}. Explain carefully how this result may be determined experimentally. Indicate the sources of error in your method and estimate the accuracy which could be achieved. [Melting-point of gallium = 30 °C.]

A ball of gallium is released from a stationary balloon, falls freely under gravity and on striking the ground it just melts. Calculate the height of the balloon assuming that the temperature of the gallium just before impact is 1 °C and that all the energy gained during its free fall is used to heat the gallium on impact. Why are the conditions specified in this problem unrealistic? [Specific latent heat of fusion of gallium = 79 kJ kg^{-1}.] (*O. and C.*)

20. Explain what is meant by the specific latent heat of vaporization of a liquid, and describe an experiment for an accurate determination of this quantity for carbon tetrachloride, which boils at 77 °C.

A thermally insulated vessel connected to a vacuum pump contains 10·0 g of water at a temperature of 0 °C. As air and water vapour are exhausted from the vessel, it is observed that the water remaining in the vessel freezes. Explain why this happens, and find the mass of water which is converted into ice. [Specific latent heat of vaporization of water at 0 °C = 2520 kJ kg^{-1}; specific latent heat of fusion of ice at 0 °C = 336 kJ kg^{-1}.] (*C.*)

21. An experiment was performed to determine the specific latent heat of vaporization of a volatile liquid at the prevailing boiling-point by the method of electrical heating. The results are summarized in the following table:

Rate of supply of energy to boiling liquid (watts)	Mass of liquid vaporized in 200 seconds (g)
10	1·6
20	6·4
30	11·2
40	16·0

Use the data to plot a graph and hence determine the specific latent heat of vaporization of the liquid and the rate of loss of heat from the calorimeter containing the boiling liquid.

Draw a labelled diagram of a suitable apparatus for use in the experiment and indicate how the above results would have been obtained. (*N.*)

3

Expansion of Gases. Kinetic Theory

Mechanical engineers, who design engines of all types, must know how the gases inside expand and contract when subjected to changes of temperature and pressure. On this account the subject of "Expansion of Gases" has considerable practical importance.

Unlike solids and liquids, the volumes of gases are considerably affected by small changes of pressure. Thus we cannot specify the condition or state of a given mass of gas without a knowledge of its pressure as well as its volume and temperature. Since the pressure, volume, and temperature of a gas must be taken into account when its expansion is studied, it is logical to examine the changes of volume with temperature *while the pressure is kept constant,* then to examine the changes of pressure *while the volume is kept constant,* and, finally, to combine the two sets of results. We therefore begin with a study of the expansion of a gas at constant pressure.

Expansion of a gas at constant pressure. Charles' law

Charles, about 1780, was among the first scientists to carry out an investigation into the expansion of a given mass of gas at constant pressure. The simple form of apparatus shown in fig. 3.1 (i) can be used for this purpose, and contains air trapped by mercury in a glass bulb B of uniform cross-section. When the level of the mercury in the open tube C is adjusted, by moving C, so that it is the same as the level of the mercury in B, the pressure of the air is atmospheric pressure (fig. 3.1 (i)). Steam is passed into a jacket round B, and when the temperature on the thermometer A is steady, C is moved so that the levels of the mercury are the same in B and C, and the length of the air column in B is read from a metre ruler S. The steam is then cut off, and the temperature of the air in B begins to decrease. A number of readings of the length of the air column in B, and the corresponding temperature, are now taken, the levels of the mercury being adjusted each time to be the same, so that the pressure of the gas is kept constant at the atmospheric pressure.

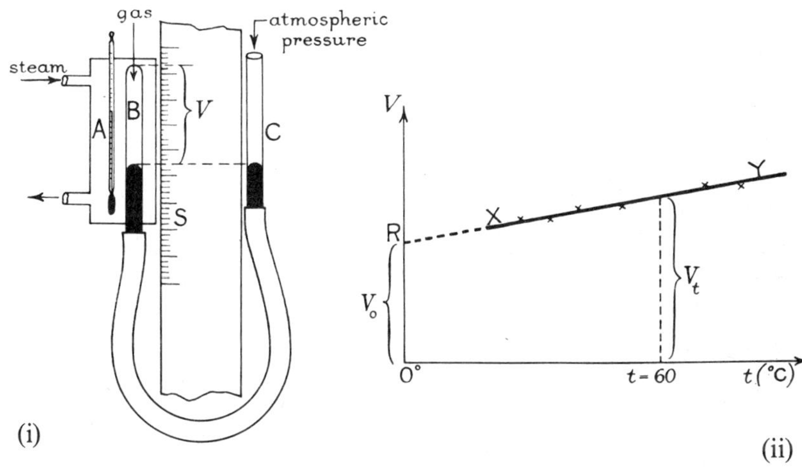

Fig. 3.1 Expansivity of gas at constant pressure

The volume V of the gas in B is proportional to the length of the air column, and experiments show that the volume-temperature graph is a straight line such as XY (fig. 3.1 (ii)). Thus the volume of the gas increases at a uniform rate for every degree rise of temperature. The same result is obtained for all other gases, and Charles found that *the volume of a given mass of gas increases by* 1/273 *of its volume at* 0 °C *for every degree Celsius rise in temperature, the pressure being constant*. This is known as *Charles' law*.

Expansivity of gas

The *expansivity* (or *volume coefficient*) of a gas may be defined as *the increase in volume per unit volume of the gas at* 0 °C *per* °C *temperature rise, the pressure being kept constant throughout*. Suppose that V_0 is the volume of a given mass of gas at 0 °C, V_t its new volume when the gas is heated to t °C, and α_p the expansivity. Then, from the latter's definition,

$$\alpha_p = \frac{\text{increase in volume from } 0\ ^\circ\text{C}}{\text{volume at } 0\ ^\circ\text{C} \times \text{temperature rise}}$$

i.e.
$$\alpha_p = \frac{V_t - V_0}{V_0 \times t} \quad . \quad . \quad . \quad . \quad . \quad . \quad (1)$$

$$\therefore V_t - V_0 = \alpha_p V_0 t$$

$$\therefore V_t = V_0(1 + \alpha_p t) \quad . \quad . \quad . \quad . \quad . \quad . \quad (2)$$

This is the formula for calculating the volume of a given mass of gas at t °C when its volume at 0 °C and its expansivity α_p are both known.

The expansivity of a gas can be found from the results of the experiment described on p. 41. The volume V_0 of the gas at 0°C is obtained by producing the straight line XY back to cut the volume axis at R, and the volume V_t at any other temperature t °C, e.g. 60 °C, is then taken. The expansivity α_p is calculated from the relation (1), and is about 1/273, or 0·003 66, per °C for all gases. In SI units, $\alpha_p = 1/273\,\text{K}^{-1}$. Thus, from (2), the volume V_t of the gas at a temperature at t °C is given by

$$V_t = V_0\left(1 + \frac{t}{273}\right) \quad . \quad . \quad . \quad . \quad . \quad . \quad (3)$$

or
$$V_t = V_0(1 + \alpha t)$$

where $\alpha_p = \alpha = 1/273$.

Absolute temperature *T*

If a given mass of gas at constant pressure is slowly cooled, its volume decreases uniformly with the temperature, as shown by the straight line YX in fig. 3.1 (ii). As the gas is cooled below 0 °C its volume shrinks further. Theoretically, then, there is a definite temperature at which the volume of the gas would become zero, although in practice the gas would liquefy before reaching this temperature.

This temperature is known as the *absolute zero,* because it is impossible to obtain any lower temperature. It can be found from the relation $V_t = V_0(1 + t/273)$ for the volume V_t of a gas at a temperature t °C. When $V_t = 0$, then $(1 + t/273) = 0$; from which it follows that $t = -273$. Thus *the absolute zero* is −273 °C, or, more accurately, −273·15 °C. 0 °C is hence 273 degrees above the absolute zero and 100 °C is 373 degrees above the absolute zero (fig. 3.2). Temperatures measured from the absolute zero are called *absolute*

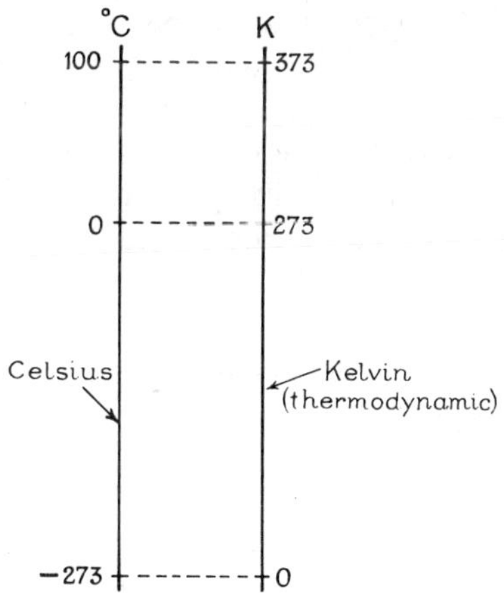

Fig. 3.2 Celsius and thermodynamic (Kelvin) scales

temperatures. Now it can be shown that absolute temperatures, using an ideal gas, are equal to the corresponding thermodynamic temperature. Hence the *kelvin* is the unit of measurement of absolute temperature. The absolute zero is 0 K, and, generally, $T(\text{K}) = 273{\cdot}15 + t(°\text{C})$ (see p. 3).

Relation between volume and absolute temperature (constant pressure)

The volume V_t of a given mass of gas is given by $V_t = V_0(1 + t/273)$.

$$\therefore\ V_t = V_0\left(\frac{273 + t}{273}\right) = \frac{V_0}{273}\,T$$

where T is the absolute temperature corresponding to t °C. But the volume V_0 at 0 °C of a given mass of gas is a constant quantity, and 273 is a constant. Consequently

$$V_t \propto T$$

so that *the volume is proportional to the absolute temperature*. Thus if the volume of a given mass of oxygen is 200 cm³ at 15 °C, its volume V at 100 °C, when the pressure is kept constant, is given by

$$\frac{V}{200} = \frac{273 + 100}{273 + 15} = \frac{373}{288}$$

$$\therefore\ V = \frac{373}{288} \times 200 = 259\ \text{cm}^3$$

We can see that the concept of absolute temperature enables the new volume of a given mass of gas to be easily calculated when its temperature is altered, the pressure being constant. The relation $V_t = V_0(1 + t/273)$ could also be used for finding the volume of a gas at 100 °C when its volume is 200 cm^3 at 15 °C, but the volume V_0 at 0 °C would first have to be calculated.

Pressure variation of gas at constant volume

We have now to turn our attention to the changes of pressure with temperature of a given mass of gas *when its volume is kept constant.* Figure 3.3 (i) illustrates a simple form of apparatus for this purpose. The gas is contained in a bulb B which is connected by narrow tubing to a mercury pressure-gauge, and the volume of the gas is kept constant by moving the open tube M until the mercury on the other side reaches a fixed mark A. The bulb is surrounded by a vessel containing water, and when the latter is heated to different constant temperatures, indicated on the thermometer D, the respective difference in levels h of the mercury is read after the volume of the gas is adjusted to be constant. If the mercury level in M is higher than the level at A, the pressure p of the gas $= (H + h)$, where H is the barometric height and h is the difference in levels (fig. 3.3 (i)). If the

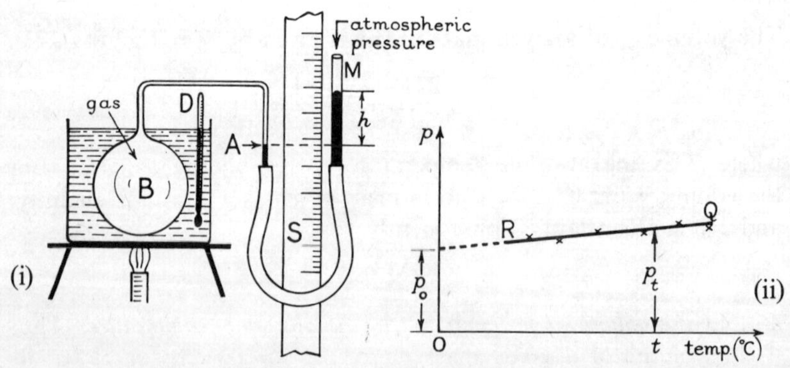

Fig. 3.3 Pressure of gas at constant volume

mercury level in M is lower than the level at A, the difference in levels is *subtracted* from the barometric height to find the corresponding pressure of the gas.

When the pressure p is plotted against the temperature t a straight-line graph RQ is obtained (fig. 3.3 (ii)). Thus experiment shows that the pressure of a given mass of gas at constant volume increases uniformly with the temperature.

Pressure coefficient of gas

The *pressure coefficient* α_V *of a given mass of gas at constant volume* is defined as *the fractional increase in pressure to its pressure at* 0 °C *per degree Celsius temperature rise.* Thus if p_0 is the pressure at 0 °C, and p_t the increased pressure at t °C, the pressure coefficient α_V is given by

$$\alpha_V = \frac{\text{increase in pressure}}{\text{pressure at } 0\ ^\circ\text{C} \times \text{temperature rise}}$$

i.e.

$$\alpha_V = \frac{p_t - p_0}{p_0 \times t} \qquad . \quad . \quad . \quad . \quad . \quad . \quad . \quad (4)$$

The pressure coefficient can be calculated from the results shown in fig. 3.3 (ii). The pressure p_0 at 0 °C is obtained by producing the line QR to intersect the pressure axis, and the pressure p_t at any other temperature t °C is then taken. The pressure coefficient is then calculated from the relation $(p_t - p_0)/p_0t$, and is found to be about 1/273 per °C for any gas, which is the same numerical value as the volume coefficient (p. 41).

From (4), it follows that the pressure p_t is given by

$$p_t = p_0(1 + \alpha_V t) = p_0\left(1 + \frac{t}{273}\right)$$

Thus

$$p_t = p_0\left(\frac{273 + t}{273}\right) = \frac{p_0}{273}T$$

where $T = 273 + t$ is the absolute temperature corresponding to t °C. Since the pressure at 0 °C of a given mass of gas at constant volume is a constant quantity, and 273 is a constant, it follows that

$$p_t \propto T \qquad . \quad . \quad . \quad . \quad . \quad . \quad . \quad (5)$$

Thus *the pressure is proportional to the absolute temperature.*

It should be carefully noted that, in the definition of the pressure coefficient, we always refer to the pressure *at* 0 °C. If a gas is heated from 40 °C to 100 °C at constant volume, and expands from a pressure p_1 to a pressure p_2, we cannot say that $p_2 = p_1[1 + (100 - 40)\alpha]$, where α is 0·003 66, because the pressure coefficient α is defined with reference to the pressure at 0 °C. The correct relation between p_2 and p_1 is deduced by noting that

$$p_2 = p_0(1 + \alpha t) = p_0(1 + 100\alpha)$$

and $$p_1 = p_0(1 + \alpha t) = p_0(1 + 40\alpha)$$

from which, by division,

$$\frac{p_2}{p_1} = \frac{1 + 100\alpha}{1 + 40\alpha}$$

Thus knowing α (0·003 66), p_2/p_1 can be evaluated. The ratio p_2/p_1 is also given, from (5), by

$$\frac{p_2}{p_1} = \frac{273 + 100}{273 + 40} = \frac{373}{313}$$

The same remarks apply to changes in volume of a gas at constant pressure. When a gas at a temperature t_1 °C and volume V_1 is heated at constant pressure to a temperature t_2 °C, its new volume V_2 can be evaluated from the relation

$$\frac{V_2}{V_1} = \frac{1 + \alpha t_2}{1 + \alpha t_1}$$

where α is 0·003 66.

Units of pressure

In SI units, pressure is measured in *newton metre*$^{-2}$, N m^{-2}. 1 N m^{-2} is also known as 1 pascal (Pa).

The pressure due to a column of liquid of height h is calculated from $p = h\rho g$. In this formula p is in N m^{-2} when h is in metres, ρ is the liquid density in kg m^{-3} and $g = 9{\cdot}8$ m s^{-2}. Thus the pressure due to a column of 760 mm mercury (Hg), density 13 600 kg m^{-3},

$$= 0{\cdot}76 \times 13\,600 \times 9{\cdot}8 = 1{\cdot}013 \times 10^5 \text{ N m}^{-2}$$

"Standard pressure", or one atmosphere, is defined as $1{\cdot}01325 \times 10^5$ N m^{-2}.

1 mmHg pressure is also called 1 torr. From $p = h\rho g$,

$$1 \text{ mmHg} = 1 \text{ torr} = 133 \text{ N m}^{-2} \text{ (Pa), approximately}$$

Boyle's law

In the seventeenth century Boyle discovered by experiment that the *pressure of a fixed mass of gas at constant temperature is inversely proportional to the volume*. This is known as *Boyle's law*. Thus if p is the pressure and V is the volume, Boyle's law states that

$$p \propto \frac{1}{V}$$

or
$$p = \frac{k}{V}$$

where k is a constant which depends on the mass of the gas, its nature, and its temperature. Consequently *the product* $pV = k$, *a constant.*

A simple apparatus to verify Boyle's law is shown in fig. 3.1 (i) on p. 41. The pressure p of the gas is varied by raising or lowering the open tube C, and the corresponding volume V of the gas is proportional to the length of B, which is read from the scale S. The temperature is kept constant during the experiment. The pressure of the gas $p = (H \pm h)$ mm of mercury, where H is the barometric height and h is the difference in levels of the mercury, the plus or minus being

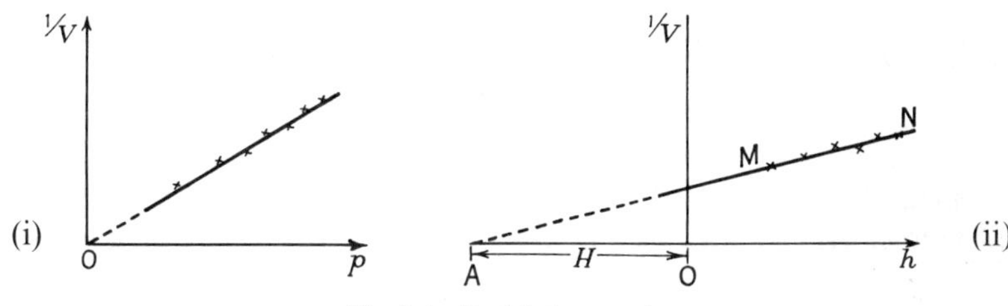

Fig. 3.4 Boyle's law graphs

used according as the level of the mercury in the open tube is above or below the level in the other tube. If a graph of $1/V$ is plotted against p, a straight line is obtained passing through the origin, as shown in fig. 3.4 (i). Thus $p \propto 1/V$, which verifies Boyle's law. We shall see later that real gases deviate from Boyle's law.

Determining barometric pressure from Boyle's law experiment

The barometric height H can be determined, if unknown, from a similar experiment to that just described. For this purpose the difference in levels h is plotted against $1/V$, when a straight line NM is obtained (fig. 3.4 (ii)).

Now the pressure p of the gas $= k/V$, assuming Boyle's law, and hence $H + h = k/V$, where H is the unknown barometric pressure. Thus when $1/V$ is zero, $H + h = 0$, i.e. $h = -H$. This means that if the line NM is produced to meet the axis of $1/V$ at A (where $1/V$ is zero), the intercept OA $= -H$. Thus the barometric pressure can be read from the horizontal axis, and is in millimetres of mercury if h is in millimetres of mercury.

To show that expansivity and pressure coefficient are equal

We have already mentioned that experiment shows that the expansivity α_p of many gases is 1/273, the same value as the pressure coefficient α_V of the gases. The equality of α_p and α_V can also be shown to follow if the gas obeys Boyle's law and Charles' law.

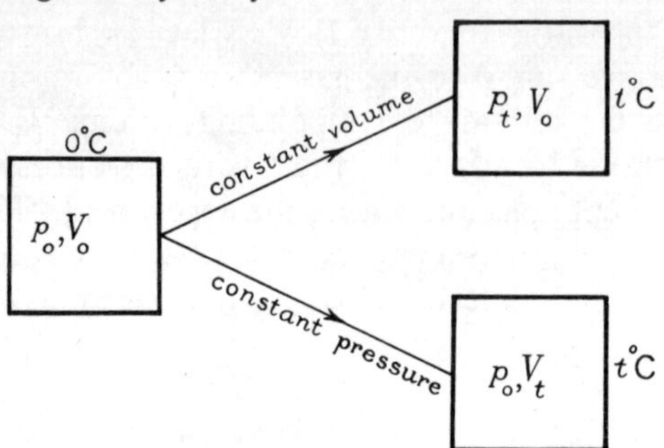

Fig. 3.5 Expansivity and pressure coefficient

Suppose that a given mass of gas at 0 °C has a pressure p_0 and a volume V_0, and that it is heated to a temperature t °C. This change of temperature can be accomplished *either* at constant volume, when the pressure changes to a value p_t, *or* at constant pressure, when the volume changes to a value V_t (see fig. 3.5). Since the temperature t °C is the same in either case, it follows from Boyle's law that

$$p_0 V_t = p_t V_0$$

$$\therefore \frac{p_t}{p_0} = \frac{V_t}{V_0} \qquad \text{(i)}$$

But $\quad V_t = V_0\left(1 + \dfrac{t}{273}\right)$ $\quad$ assuming Charles' law.

$$\therefore \frac{V_t}{V_0} = 1 + \frac{t}{273}$$

Substituting in (i) for V_t/V_0, we have

$$\frac{p_t}{p_0} = 1 + \frac{t}{273}$$

Thus $\quad p_t = p_0\left(1 + \dfrac{t}{273}\right)$

from which it follows that the pressure coefficient α_V is 1/273. Thus the expansivity of the gas is equal to its pressure coefficient if it obeys Boyle's and Charles' laws, the property of an *ideal* gas.

The ideal gas equation

Consider a *given mass* of an ideal gas with a pressure p_1, a volume V_1, and an absolute temperature T_1, and suppose it is heated so that its pressure changes to p_2, its volume to V_2, and its absolute temperature to T_2. The change from p_1, V_1, T_1, to p_2, V_2, T_2 can be made in two stages, as represented in fig. 3.6. (i) Keep the pressure constant at the value p_1 and heat the gas until its absolute temperature is T_2; (ii) then keep the temperature constant at the absolute value T_2, and change the pressure from p_1 to p_2, when the volume of the gas changes to a value denoted by V_2. The final condition of the gas is now represented by p_2, V_2, T_2.

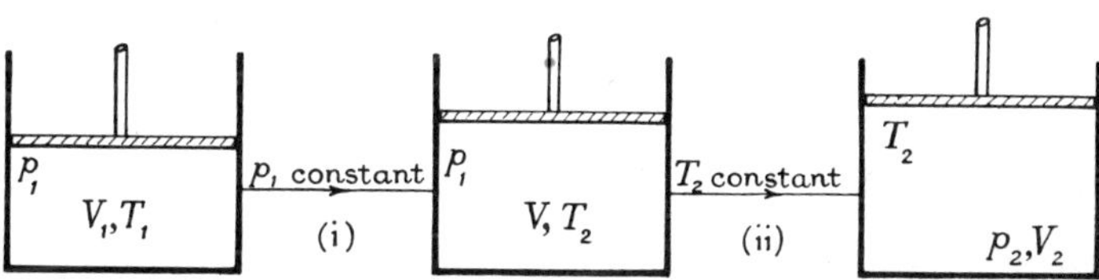

Fig. 3.6 Ideal gas equation

Stage (i).—Let V be the volume of the gas when p_1 is kept constant and the absolute temperature changes from T_1 to T_2. Then, since volume $\propto T$ (p. 43), we have

$$\frac{V}{V_1} = \frac{T_2}{T_1}$$

i.e.
$$V = \frac{T_2 V_1}{T_1} \qquad \text{(i)}$$

Stage (ii).—When the temperature is kept constant and the pressure and volume of the gas change, Boyle's law is obeyed,

$$\therefore p_1 V = p_2 V_2$$

i.e.
$$V = \frac{p_2 V_2}{p_1} \qquad \text{(ii)}$$

From (i) and (ii), it follows that

$$\frac{p_2 V_2}{p_1} = \frac{T_2 V_1}{T_1}$$

$$\therefore \frac{p_2 V_2}{T_2} = \frac{p_1 V_1}{T_1}$$

If the given mass of gas is heated again so that it changes from a condition represented by p_2, V_2, T_2 to a condition represented by p_0, V_0, T_0, then, from the above reasoning,

$$\frac{p_2 V_2}{T_2} = \frac{p_0 V_0}{T_0}$$

It can now be seen that for a *given mass* of gas,

$$\frac{pV}{T} = K$$

where p, V, T denote the pressure, volume, and absolute temperature of the gas, and K *is a constant depending on the nature and mass of the gas*. Thus, generally,

$$pV = KT \quad . \quad . \quad . \quad . \quad . \quad . \quad . \quad . \quad (6)$$

Magnitude of gas constant

If we choose to deal with unit mass of a given gas (such as 1 kg) the corresponding gas constant will be denoted by the symbol R.

Thus

$$pV = RT \quad . \quad . \quad . \quad . \quad . \quad . \quad . \quad . \quad (7)$$

for unit mass of a gas, and this equation is known as the *characteristic equation of an ideal gas*.

The gas constant R for oxygen can be calculated from the observation that the volume of 1 kg of the gas at s.t.p. is $0{\cdot}7\ \mathrm{m}^3$. The pressure p of the gas is 760 mm of mercury, and since the pressure is calculated from the formula $p = h\rho g$, then

$$p = 0{\cdot}76 \times 13\,600 \times 9{\cdot}8\ \mathrm{N\,m^{-2}}$$

The temperature of the gas is 0 °C, so that its absolute temperature $T = 273$ K.

$$\therefore\ R = \frac{pV}{T} = \frac{0{\cdot}76 \times 13\,600 \times 9{\cdot}8 \times 0{\cdot}7}{273} = 260$$

As we shall show on p. 98, the product pV is a measure of work or energy, and hence R $(= pV/T)$ is expressed in "joules per kg per kelvin" or $\mathrm{J\,kg^{-1}\,K^{-1}}$. Thus the gas constant for 1 kg of oxygen is $260\ \mathrm{J\,kg^{-1}\,K^{-1}}$ or $0{\cdot}26\ \mathrm{kJ\,kg^{-1}\,K^{-1}}$.

The numerical value of R will be different for one kg of different gases. In the case of hydrogen, for example, 1 kg occupies a volume of $11{\cdot}2\ \text{m}^3$ at s.t.p., so that $V = 11{\cdot}2\ \text{m}^3$, $p = 0{\cdot}76 \times 13\,600 \times 9{\cdot}8\ \text{N m}^{-2}$ and $T = 273$ K. The gas constant for hydrogen is thus given by

$$R = \frac{pV}{T} = \frac{0{\cdot}76 \times 13\,600 \times 9{\cdot}8 \times 11{\cdot}2}{273}$$

$$= 4150\ \text{J kg}^{-1}\ \text{K}^{-1}$$

$$= 4{\cdot}15\ \text{kJ kg}^{-1}\ \text{K}^{-1}$$

General gas equation

We have now to consider the gas equation for m kg of a given gas. Suppose, for example, that we have 5 kg of oxygen. At s.t.p. its volume will be 5 times as great as the volume of 1 kg at s.t.p., so that the volume $= 5 \times 0{\cdot}7\ \text{m}^3$ (p. 50). Now the gas constant K for 5 kg of oxygen is given by

$$K = \frac{pV}{T}$$

where p is standard atmospheric pressure, $T = 273$ K, and $V = 5 \times 0{\cdot}7\ \text{m}^3$. Since the same values of p and T are obtained for 1 kg of oxygen at s.t.p., but the volume is then $0{\cdot}7\ \text{m}^3$ instead of $5 \times 0{\cdot}7\ \text{m}^3$, it follows that K is five times as large as the gas constant R for 1 kg of the gas. Hence $K = 5R$, and thus $5R = pV/T$, or $pV = 5RT$ for 5 kg of oxygen. If there are m kg of a given gas, it can be seen that the gas equation can be written as

$$pV = mRT \qquad \text{(8)}$$

R being the gas constant for 1 kg.

From this equation, the mass m of a gas is given by

$$m = \frac{pV}{RT}$$

This equation can be used to find the mass of a volume of gas, provided the gas constant R for 1 kg is known.

As an illustration, suppose that the volume of a quantity of oxygen collected at 15 °C and 750 mmHg pressure is 1·2 litres or $1{\cdot}2 \times 10^{-3}\ \text{m}^3$. Then, using $R = 260\ \text{J kg}^{-1}\ \text{K}^{-1}$ for oxygen, the mass of the gas is given by

$$m = \frac{pV}{RT} = \frac{(0{\cdot}75 \times 13\,600 \times 9{\cdot}8) \times 1{\cdot}2 \times 10^{-3}}{260 \times 288}$$

$$= 1{\cdot}6 \times 10^{-3}\ \text{kg} = 1{\cdot}6\ \text{g}$$

Molar gas constant

From Avogadro's hypothesis, the volume occupied by a mole of *any* gas at s.t.p. is the same and is found to equal 22·4 litres ($22{\cdot}4 \times 10^{-3}$ m^3) approximately. Since $pV = KT$, it follows that *the gas constant for one mole of any gas is the same*, as p, V, T have the same values in all cases.

The *molar gas constant*, which we shall denote by **R**, is given by

$$\mathbf{R} = \frac{pV}{T} = \frac{(0{\cdot}76 \times 13\,600 \times 9{\cdot}8) \times 22{\cdot}4 \times 10^{-3}}{273}$$

$$= 8{\cdot}3 \text{ J mol}^{-1}\text{ K}^{-1} = 8{\cdot}3 \text{ kJ kmol}^{-1}\text{ K}^{-1}$$

where 1 kmol = 1000 mol. The molar gas equation is $pV = \mathbf{R}T$. It is related to the gas equation per unit mass of the same gas, $pV = RT$ by the fact that

$$\mathbf{R} = mR$$

where m is the molar mass. Thus the gas constant per kilogramme of hydrogen, for example, is $R = 4{\cdot}15$ kJ kg^{-1} K^{-1} = 4·15 J g^{-1} K^{-1}. Thus

$$\text{molar mass } m = \frac{\mathbf{R}}{R} = \frac{8{\cdot}3}{4{\cdot}15} = 2\text{g}$$

Similarly, the value of R for oxygen is 0·26 kJ kg^{-1} K^{-1} or 0·26 J g^{-1} K^{-1}.

$$\therefore \text{ molar mass of oxygen} = \frac{\mathbf{R}}{R} = \frac{8{\cdot}3}{0{\cdot}26}\text{ g} = 32\text{ g}$$

Examples

1. Explain the difference between the Celsius and the absolute scales of temperature.
It is found that the volume of a certain gas increases in the ratio 1·035:1 between 15 °C and 25 °C. Calculate the absolute zero on the Celsius scale for this gas. (*O. and C.*)

Suppose the absolute zero is x °C below 0 °C. Then, assuming the change of volume takes place at constant pressure, the volume is proportional to the absolute temperature. Hence

$$\frac{x + 25}{x + 15} = \frac{1{\cdot}035}{1}$$

from which $x = -270{\cdot}7$ °C

2. Explain what is meant by cubic expansivity. Describe an experiment to determine the cubic expansivity of a fixed mass of air kept at constant pressure.

The density of argon is 1·60 kg m^{-3} at 27 °C and at a pressure of 750 mm of mercury.

What is the mass of argon in an argon-filled electric lamp bulb of volume 100 cm³ if the pressure inside is 750 mm of mercury when the average temperature of the gas is 120°C? (*N.*)

The volume of a fixed mass of gas at a constant pressure is proportional to its absolute temperature. Thus the volume V at 27 °C of 100 cm³ of argon at 120 °C and 750 mm pressure is given by

$$\frac{V}{100} = \frac{273 + 27}{273 + 120}$$

$$\text{i.e. } V = \frac{300}{393} \times 100 = 76{\cdot}3 \text{ cm}^3$$

the pressure being constant at 750 mmHg.

But 10^6 cm³ of argon at 27 °C and 750 mm pressure has mass 1·60 kg.

∴ 76·3 cm³ of argon at 27 °C and 750 mm pressure has mass $\frac{76{\cdot}3}{10^6} \times 1{\cdot}60$ kg

$$= 1{\cdot}22 \times 10^{-4} \text{ kg}$$

This is the mass of argon in the electric lamp.

3. Describe how you would determine the temperature of a room by means of a constant-volume air thermometer.

A diving-bell of uniform cross-section and 2 metres high is full of air at 70 °C when the atmospheric pressure is equal to 77 cm of mercury. It is then sunk into water at 27 °C and the water rises 50 cm within the bell. Assuming that no air escapes as the bell is sunk, find the depth of the bell below the water surface. [Take the density of mercury as 13 600 kg m⁻³.] (*L.*)

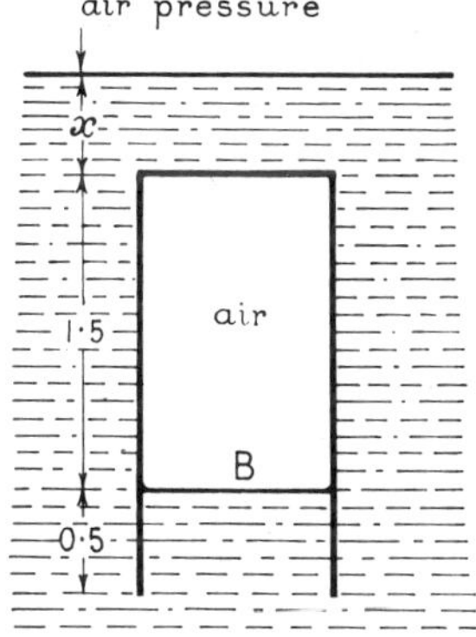

Fig. 3.7 Example

First part.—The temperature of a room in °C is given by

$$t = \frac{p_t - p_{ice}}{p_{st} - p_{ice}} \times 100$$

where p_t is the pressure at room temperature, and p_{st}, p_0 are the respective pressures at the temperature of steam at normal atmospheric pressure and at the melting-point of ice (see p. 4).

Second part.—When the level of the water rises 50 cm to B (fig. 3.7), the volume V_1 of air inside the bell = 1·5 A m³, where A is the uniform cross-section of the bell in m².

The pressure p_1 of the air = the pressure at B

$$= 0{\cdot}77 \times 13\,600\, g + (1{\cdot}5 + x)\, 1000\, g \text{ N m}^{-2}$$

where x m is the depth of the top of the bell below the surface of the water and $g = 9{\cdot}8$. The absolute temperature T_1 of the air = 273 + 27 = 300 K.

Now the original pressure p_2 of the air in the bell = 0·77 × 13 600 g N m⁻², and the original volume V_2 of the air in the bell = 2 A m³. The original absolute temperature = 273 + 70 = 343 K.

Since the mass of the air in the bell is constant,

$$\frac{p_1 V_1}{T_1} = \frac{p_2 V_2}{T_2}$$

$$\therefore \quad \frac{(0{\cdot}77 \times 13\,600 + 1500 + 1000\,x)g \times 1{\cdot}5\,A}{300} = \frac{0{\cdot}77 \times 13\,600\,g \times 2\,A}{343}$$

$$\therefore 0{\cdot}77 \times 13\,600 + 1500 + 1000\,x = \frac{0{\cdot}77 \times 13\,600 \times 2 \times 300}{1{\cdot}5 \times 343}$$

$$\therefore x = 0{\cdot}24 \text{ m} = 24{\cdot}0 \text{ cm}.$$

4. State the laws showing the relation between pressure, volume, and temperature of a fixed mass of gas. Explain the experimental method of verifying one of the laws.

Two glass bulbs of volume 25 and 35 cm^3 are connected by a narrow tube of negligible volume. The apparatus is filled with air at 0 °C and 760 mm mercury pressure and sealed off. If the 25 cm^3 bulb is maintained at 0 °C and the 35 cm^3 bulb at 100 °C, what is the new pressure within the apparatus? (*L.*)

At 0 °C and 760 mm pressure, the total volume of air = 25 + 35 = 60 cm^3 = $60 \times 10^{-6}\,m^3$. Since the mass m of a gas is given by $m = pV/RT$, the total mass of air in the two bulbs $= 0{\cdot}76 \times 13\,600 \times 9{\cdot}8 \times 60 \times 10^{-6}/(R \times 273)$, where R is the gas constant per kg of air (see p. 50).

The temperature of the 35 cm^3 bulb is changed to 100 °C, or 373 K. Hence the mass of air in this bulb $= p \times 35 \times 10^{-6}/(R \times 373)$, where p is the new pressure in the apparatus. Similarly, the mass of air now present in the 25 cm^3 bulb $= p \times 25 \times 10^{-6}/(R \times 273)$. *But the total mass of air is constant.* Hence, from above, cancelling 10^{-6},

$$\frac{p \times 35}{R \times 373} + \frac{p \times 25}{R \times 273} = \frac{0{\cdot}76 \times 13\,600 \times 9{\cdot}8 \times 60}{R \times 273}$$

Dividing throughout by $1/R$,

$$\therefore \frac{35p}{373} + \frac{25p}{273} = \frac{0{\cdot}76 \times 13\,600 \times 9{\cdot}8 \times 60}{273}$$

from which $p = 901$ mmHg

Intermolecular energy and forces

Every substance, whether solid, liquid, or gaseous, is made up of many millions of elementary particles known as *molecules* of the substance. John Dalton, in 1803, was the first person to make use of this conception of matter. Later, it was realized that the molecules of a substance had some form of movement, and there developed the *Kinetic Theory of Matter*. In the hands of Clerk Maxwell, this enabled the properties of a gas to be explained, as we see shortly.

The intermolecular *potential energy* V for two molecules varies with their distance of separation r as shown by the curve in fig. 3.8 (i). The minimum potential energy occurs at X, at a distance $r = r_0$, and when the temperature is so low that the kinetic energy is very small, the total energy at X is a minimum. Thus $r = r_0$ corresponds to the equilibrium separation of the molecules. For most solids, r_0 is of the order of 2 or 3×10^{-10} m.

The intermolecular *force* F between the molecules is given by

$F = -dV/dr$. The variation of F with r thus follows the negative gradient of the $V-r$ curve and is shown in fig. 3.8 (ii). For intermolecular distances greater than r_0, F is an *attractive* force and follows an inverse-power law of r when r is much greater than r_0. For intermolecular distances less than r_0, F is a *repulsive* force. At X, the attractive and repulsive forces balance each other.

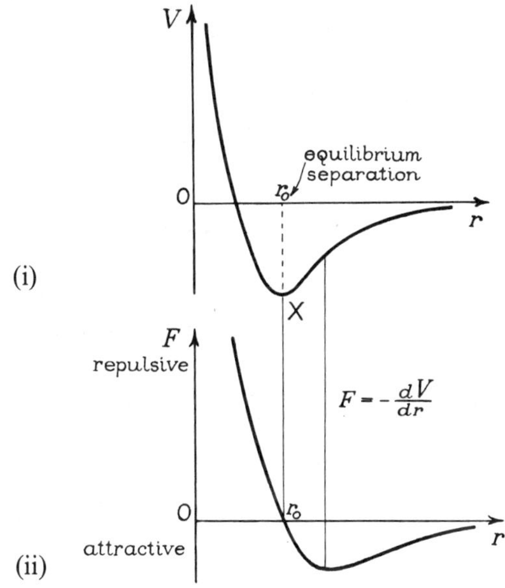

Fig. 3.8 Intermolecular potential energy and force

The graphs in fig. 3.8 apply to all phases of matter, whether gaseous, liquid or solid. In the case of a gas, the molecules are so far apart from each other on the average that there is little intermolecular attraction. The cohesive force between the particles of matter which constitute a gas is thus extremely small, and this can be seen by the fact that a gas will automatically fill any space into which it is led. The intermolecular attraction between the particles of a *liquid* is greater than that which exists between the molecules of a gas because they are closer together on the average. Thus, although a liquid always assumes the shape of the vessel in which it is placed, a given amount of it occupies a definite volume, unlike a gas. The bonds between the molecules of a *solid* are stronger than those between the molecules of a liquid, as the solid has a fixed shape. Thus all solids have a more ordered structure than their liquid phase, that is, the atoms and molecules are in a definite pattern or lattice. This makes a very strong structure, whereas, in a liquid, the molecules are only partially ordered in structure. On this account solids are less compressible than liquids, that is, solids have higher elastic modulus values than their corresponding liquids.

Elementary kinetic theory of gases. Assumptions

According to the kinetic theory, the molecules of a gas move about in all different directions, making collisions for a short time with each other and with the walls of the containing vessel, and rebounding from them. The motion of the molecules is random or haphazard. The average impacts per second of the molecules on one side of the container creates a particular pressure there, which is the pressure of the gas. When the gas is heated the molecules move faster, and hence their average kinetic energy, and pressure, increase.

In 1860 Clerk Maxwell showed theoretically that the well-known properties of a gas were capable of being explained on the basis of the following assumptions:

(1) The molecules of the gas behaved like elastic spheres.

(2) The time of collision between the molecules was negligible.

(3) The attraction between the molecules was negligible.

(4) The volume of the molecules was negligible compared with the volume occupied by the gas.

Maxwell used advanced mathematics in his statistical treatment of the "Kinetic Theory of Gases", as the subject is called, and this is beyond the scope of the book. The following elementary treatment, while not perfectly rigid, will enable us to reach Maxwell's main result.

Pressure calculation

Consider a cube of length l containing a gas of n molecules each of mass m. The molecules are moving randomly, i.e. in all different directions. As a simplification, we can imagine that $\frac{1}{3}n$ molecules are moving in each of three perpendicular directions, Ox, Oy, Oz, which we shall suppose are parallel to the faces of the cube (fig. 3.9). Then the number of molecules moving in *one* direction, e.g. perpendicular to the face A, is $\frac{1}{2}$ of $\frac{1}{3}n$, or $\frac{1}{6}n$.

Suppose that a particular molecule moves with a translational velocity u along Ox towards the face A. The momentum of the molecule is mu, since m is its mass. Momentum is a vector quantity;

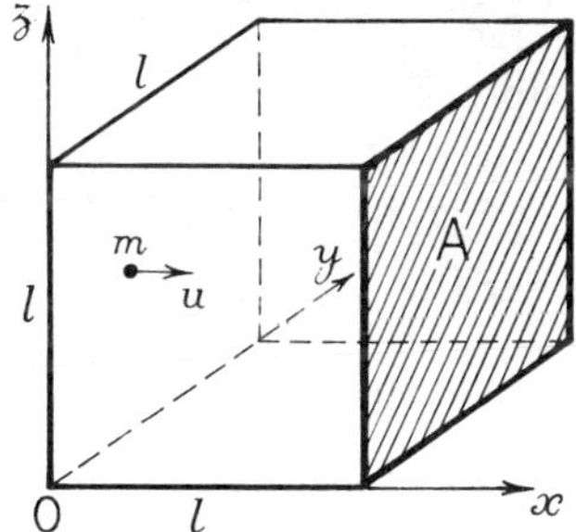

Fig. 3.9 Kinetic theory of gases—pressure proportional to mean square velocity

it has direction as well as magnitude. Thus the momentum on rebounding is $-mu$ if the collision is perfectly elastic. Hence the change in momentum $= mu-(-mu) = 2mu$. The molecule takes a time l/u to travel from one side to the other of the cube, and hence the number of impacts *per second* on A of each of the $\frac{1}{6}n$ molecules is u/l.

$$\therefore \text{ total change in momentum per second } = \tfrac{1}{6}n \times (u/l) \times 2mu$$
$$= \tfrac{1}{3}nmu^2/l$$

But force exerted on A = total change in momentum per second.

$$\therefore \text{ pressure } p \text{ on A} = \frac{\text{force}}{\text{area}} = (\tfrac{1}{3}nmu^2/l) \div l^2 = \tfrac{1}{3}nmu^2/l^3$$

But volume of the gas $V = l^3$

$$\therefore pV = \tfrac{1}{3}nmu^2$$

Thus the pressure depends on the *square* of the velocity. Now the molecules are all moving with different velocities (p. 63). We therefore replace u^2 by the *average of the squares of the velocities of all the molecules*, which will be denoted by $\overline{c^2}$. Thus

$$pV = \tfrac{1}{3}nm\overline{c^2} \qquad \text{(i)}$$

This expression was deduced by Maxwell by a more rigid mathematical calculation (p. 63).

The physical reason for the appearance of the square of the velocity in the expression for pressure should be observed by the reader. The change of momentum on impact is proportional to the velocity. The number of impacts per second on a given wall is also proportional to the velocity. Thus the rate of change of momentum, or force, is proportional to velocity × velocity or (velocity)^2.

Although we have considered a gas molecule to make no collisions in travelling from one face to another of the cube, collisions occur for real gases. A collision between a fast- and slow-moving molecule will generally cause the fast molecule to slow down and the slow molecule to speed up. At any given instant, then, all the molecules have different speeds (see p. 63). It should be noted, however, that if collisions are elastic, there is no loss of momentum (or energy) on collision. Thus the momentum with which a molecule strikes a wall may be considered to be some average value which is independent of collisions.

Numerical example of pressure

To illustrate numerically the calculation of the pressure, consider a hollow cube of side 10 cm containing 10^{22} oxygen molecules at a constant temperature. Suppose the average translational speed of the molecules is 500 m s^{-1} and the mass of each molecule is 5×10^{-26} kg. Then, for one of the molecules moving perpendicularly to a side of the cube,

$$\text{momentum change on impact with side} = 2\,mu$$

$$= 2 \times 5 \times 10^{-26} \times 500 \text{ kg m s}^{-1}$$

$$= 5 \times 10^{-23} \text{ kg m s}^{-1}$$

Time to make successive impacts on same side = time to travel 2×10 cm or 2×10^{-1}m.

$$\therefore \text{time} = \frac{2 \times 10^{-1}}{500} = 4 \times 10^{-4}\text{s}$$

$$\therefore \text{momentum change per second} = \frac{5 \times 10^{-23}}{4 \times 10^{-4}} = 1{\cdot}25 \times 10^{-19} \text{ N}$$

$$\therefore \text{force on side assumed due to one-third of molecules}$$

$$= 1{\cdot}25 \times 10^{-19} \times \tfrac{1}{3} \times 10^{22} \text{ N}$$

$$= 400 \text{ N (approx)}$$

$$\therefore \text{pressure} = \frac{\text{force}}{\text{area}} = \frac{400 \text{ N}}{100 \times 10^{-4} \text{ m}^2} = 4 \times 10^4 \text{ N m}^{-2}$$

Now $$760 \text{ mmHg pressure} = 1{\cdot}01325 \times 10^5 \text{ N m}^{-2}$$

$$\therefore \text{gas pressure} = \frac{4 \times 10^4}{10^5} \times 760 \text{ mmHg (approx)} = 300 \text{ mmHg (approx)}$$

A cube of side 10 cm has a volume of 1000 cm^3 or 1 litre. About 6×10^{23} molecules, 60 times 10^{22} molecules, of oxygen occupy a volume of about 22 litres at 0 °C and 760 mmHg pressure. The lower pressure of 300 mmHg is due mainly to the relatively fewer molecules in the cube or the small mass of gas present.

Translational kinetic energy of gas

Since $nm = M$, where M is the mass of the gas, the average translational kinetic energy of the gas $E = \frac{1}{2}M\overline{c^2} = \frac{1}{2}nm\overline{c^2}$. Thus, from (i),

$$pV = \tfrac{1}{3}nm\overline{c^2} = \tfrac{1}{3}M\overline{c^2} = \tfrac{2}{3}E$$

We now assume that the mean translational kinetic energy E of a given mass of gas depends only on its temperature, in accordance with the theory that heat is a form of energy (p. 13). With this assumption, *pV is a constant at a given temperature*, which is Boyle's law.

Experiment shows that $pV = KT$, where K is the gas constant for the given mass of gas. Consequently, from above, $\frac{2}{3}E = KT$, and hence $E \propto T$. Thus *the mean translational kinetic energy of a gas is directly proportional to its absolute temperature.*

It should be noted here that the molecules of the gas are assumed to have only translational kinetic energy. This is the case where molecules are monatomic. Diatomic and more complex molecules, however, have rotational and vibrational kinetic energy, in addition to translational kinetic energy, and this is considered later.

Root-mean-square (r.m.s.) velocity of a gas

The quantity $\overline{c^2}$, the "average of the squares of the velocities" of the molecules, can be understood by supposing that we have four molecules with velocities 1100, 1000, 1050, 1070 metres per second respectively. The squares of their velocities are then numerically 1100^2, 1000^2, 1050^2, 1070^2, and the average of the squares of the velocities $= \frac{1}{4}(1100^2 + 1000^2 + 1050^2 + 1070^2)$. The square root of this average is known as the *root-mean-square* (*r.m.s.*) value of the four velocities, the term "root" implying "square root" and the term "mean" implying "average". If we consider the n molecules of the gas, with respective velocities $u_1, u_2, u_3, \ldots, u_n$,

the root-mean-square velocity

$$= \sqrt{\frac{1}{n}({u_1}^2 + {u_2}^2 + {u_3}^3 + \ldots + {u_n}^2)}$$

This should not be confused with the *average velocity* of the molecules, which is given by

$$\frac{1}{n}(u_1 + u_2 + u_3 + \ldots + u_n)$$

The average velocity has not the same value as the r.m.s. velocity, as the reader can test by taking four numbers and calculating the two quantities (see also p. 63).

As we have used the symbol $\overline{c^2}$ for the average value of the squares of the velocities, the r.m.s. value of the velocities is denoted by $\sqrt{\overline{c^2}}$.

The r.m.s. value can be calculated from the relation

$$pV = \tfrac{1}{3}nm\overline{c^2} = \tfrac{1}{3}M\overline{c^2}$$

from which

$$\overline{c^2} = \frac{3pV}{M} = \frac{3p}{\rho}$$

where $\rho = \dfrac{M}{V}$ = the density of the gas.

$$\therefore \text{r.m.s. value} = \sqrt{\overline{c^2}} = \sqrt{\frac{3p}{\rho}}$$

In this expression, the r.m.s. velocity is in m s^{-1} (metre second^{-1}) when p is in N m^{-2} and ρ is in kg m^{-3}. The velocity of sound in a gas is given by a similar expression, $v = \sqrt{(\gamma p/\rho)}$, where γ is 1·4 for air, for example. This is due to the fact that the translational movement of the molecules controls the speed of movement of the sound wave in the gas.

r.m.s. velocity of gases. Earth's atmosphere

The r.m.s. velocity of hydrogen molecules can now be calculated. The density of hydrogen at 0 °C and $1{\cdot}013 \times 10^5$ N m^{-2} (standard) pressure is about 0·09 kg m^{-3}.

$$\therefore \sqrt{\overline{c^2}} = \sqrt{\frac{3p}{\rho}} = \sqrt{\frac{3 \times 1{\cdot}013 \times 10^5}{0{\cdot}09}}$$

$$= 1840 \text{ m s}^{-1}$$

The r.m.s. velocity of hydrogen molecules is thus nearly 2 km per second, or about 7000 km per hour. The r.m.s. velocities of molecules of other gases are less than that of hydrogen, since r.m.s. velocity $\propto 1/\sqrt{\rho}$. Oxygen gas, which is about 16 times as dense as hydrogen, has thus $1/\sqrt{16}$ or 1/4 of the r.m.s. velocity, about 460 m s^{-1}. Air has a density of 1·29 kg m^{-3}, about 14·4 times that of hydrogen. The r.m.s. velocity of air molecules is hence $(1/\sqrt{14{\cdot}4}) \times 1840$ m s^{-1}, about 500 m s^{-1}.

The *escape velocity* of an object launched from the earth, the least velocity which can overcome completely the earth's gravitational attraction, is about 11 km s^{-1}. The earth's atmosphere stays round the earth because most air molecules have speeds less than the escape velocity of this planet. Only small quantities of helium are found close to the earth because it is an extremely light gas and diffuses upwards (p. 61). The moon has little or no atmosphere. The gravitational attraction is about one-sixth that of the earth, so the escape velocity is low.

Graham's law, Dalton's law, Avogadro's hypothesis

Graham found that *the rate of diffusion of a gas through a porous partition was inversely proportional to the square root of its density.* This is known as *Graham's law of diffusion.*

On the kinetic theory of gases, the rate of diffusion through a fine hole is proportional to the average or mean velocity $\bar{c}$ of the gas. The mean velocity $\bar{c}$ must not be confused with the root-mean-square (r.m.s.) velocity $\sqrt{\overline{c^2}}$, as explained on p. 59. Maxwell has shown that $\bar{c} = 0{\cdot}92\sqrt{\overline{c^2}}$ (approx), and, generally, $\bar{c} \propto \sqrt{\overline{c^2}}$.

$$\therefore \text{ rate of diffusion} \propto \sqrt{\overline{c^2}}$$

But, from p. 60,

$$\sqrt{\overline{c^2}} = \sqrt{\frac{3p}{\rho}}$$

where p is the pressure and ρ is the density. Thus, for a given gas pressure,

$$\text{rate of diffusion} \propto \frac{1}{\sqrt{\rho}}$$

This explains Graham's law.

Dalton's law of partial pressures states that the total pressure of a mixture of gases is equal to the sum of the pressures of the individual gases, assuming each gas occupies the volume of the mixture at the same temperature (p. 131). On the kinetic theory, the total pressure p in a volume V due to the mixture is given by

$$pV = \tfrac{1}{3}n_1m_1\overline{c_1{}^2} + \tfrac{1}{3}n_2m_2\overline{c_2{}^2} + \dots$$

by similar reasoning to that on p. 57. This is an expression of Dalton's law, since, if they occupy the same volume as the mixture, each gas has a pressure given respectively by

$$p_1V = \tfrac{1}{3}n_1m_1\overline{c_1{}^2};\, p_2V = \tfrac{1}{3}n_2m_2\overline{c_2{}^2} \dots .$$

Avogadro's hypothesis states that *at the same temperature and pressure, equal volumes of gases contain the same number of molecules.* On the kinetic theory, since $pV = \tfrac{1}{3}nm\overline{c^2}$, it follows that for two different gases 1 and 2 at the same pressure and volume,

$$pV = \tfrac{1}{3}n_1m_1\overline{c_1{}^2} = \tfrac{1}{3}n_2m_2\overline{c_2{}^2} \quad \dots \quad \text{(i)}$$

If the two gases are mixed, then since their temperatures are equal, we do not obtain an exchange of kinetic energy between individual molecules. Thus, as Maxwell showed rigidly,

$$\tfrac{1}{2}m_1\overline{c_1^2} = \tfrac{1}{2}m_2\overline{c_2^2} \quad . \quad . \quad . \quad . \quad . \quad . \quad \text{(ii)}$$

Hence, from (i) and (ii), $n_1 = n_2$, that is, the numbers of molecules in equal volumes of gases are equal.

Measurement of plasma temperature

The temperature of very hot ionized gases or *plasma*, obtained when heavy electrical discharges are passed through the gas, can be measured from the Doppler effect produced. Owing to their random velocities, some molecules are moving away from the observer at any instant and others are moving towards him. In the former case, the apparent wavelength of the light emitted is increased; in the latter case, the apparent wavelength is decreased. The observer thus sees a broadening of the spectral line, and the breadth of the line is a measure of the temperature of the gas, as we shall now show.

Suppose λ is the wavelength of the light emitted by a stationary source. Then $f = c/\lambda$, where f is the frequency of the light waves and c is the velocity of light. If the source moves directly away from the observer with a velocity v, the f waves emitted per second now occupy a distance $(c + v)$.

$\therefore$ apparent wavelength λ'

$$= \frac{c+v}{f} = \frac{c+v}{c}\lambda = \left(1 + \frac{v}{c}\right)\lambda \quad . \quad . \quad . \quad . \quad . \quad . \quad \text{(i)}$$

Similarly, if the source of light moves towards the observer with a velocity v, then

apparent wavelength λ''

$$= \frac{c-v}{f} = \frac{c-v}{c}\lambda = \left(1 - \frac{v}{c}\right)\lambda \quad . \quad . \quad . \quad . \quad . \quad . \quad \text{(ii)}$$

$$\therefore \text{breadth of spectral line} = \lambda' - \lambda'' = \frac{2v}{c}\lambda \quad . \quad . \quad . \quad . \quad \text{(iii)}$$

The breadth is the change in wavelength corresponding to the edges of the line, and can be measured by a diffraction grating. Then, knowing c and λ, the mean velocity v of the molecules can be calculated from (iii). The r.m.s. velocity is about 1·05 times the mean velocity, and is equal to $\sqrt{(3RT)}$, where T is the absolute temperature of the gas and R is the gas constant per unit mass. Thus T can be calculated.

Maxwell's distribution law

Clerk Maxwell showed that the velocities of the molecules of a gas in an enclosure are distributed according to a law expressed by

$$\delta N = NAc^2e^{-\beta c^2}.\delta c \quad . \quad . \quad . \quad . \quad . \quad \text{(i)}$$

where δN is the number of molecules having velocities in the range c to $c + \delta c$, N is the total number of molecules, and A, β are constants given by $A = (2m^3/\pi k^3 T^3)^{1/2}$, $\beta = m/2kT$. Here m is the mass of a molecule, T is the gas absolute temperature and k is *Boltzmann's constant.* $k = \mathbf{R}/N_A$, where $\mathbf{R}$ is the molar gas constant and N_A is the number of molecules in one mole of any gas, called *Avogadro's constant.* k has the value $1{\cdot}38 \times 10^{-23}$ J K^{-1}. Thus A and β in equation (i) depend on the nature and temperature of the gas.

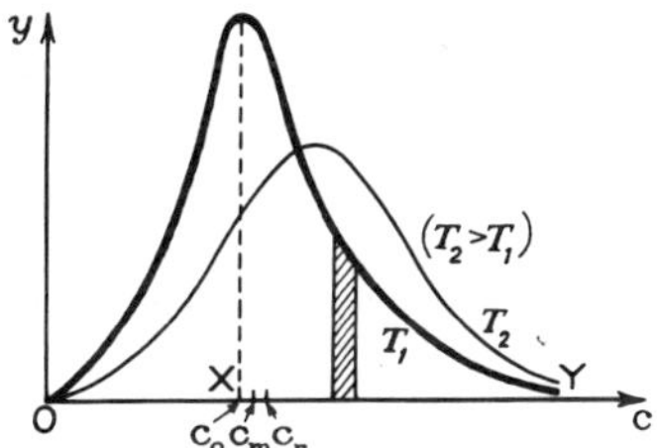

Fig. 3.10 Distribution of molecular speeds

If the function $y = Ac^2e^{-\beta c^2}$ is plotted against c for a particular temperature T_1 (fig. 3.10), the area $y.\delta c$ of the shaded strip is then a measure of $\delta N/N$, from (i). This is the fractional number of molecules which have a velocity in the range c to $c + \delta c$. The total area between the curve and the axis of c thus represents unity. The velocity c_0 corresponding to the peak of the curve is known as the *most probable velocity*, because there are more molecules in the range c_0 to $c_0 + \delta c$ than at any other velocity and the same value of δc. The *mean velocity* c_m is the average velocity of all the molecules. The area between the curve and XY to the right of the peak is greater than the area above XO, and hence the number of molecules with a velocity greater than c_0 is greater than the number with a velocity less than c_0. Consequently the mean velocity c_m is greater than c_0. The *root-mean-square (r.m.s.) velocity* c_r is the square root of the mean of the squares of the velocities of all the molecules. From Maxwell's distribution law given in (i), it can be shown that

$$c_0 : c_m : c_r = 1{\cdot}00 : 1{\cdot}13 : 1{\cdot}23 \quad . \quad . \quad . \quad . \quad . \quad . \quad . \quad . \quad \text{(ii)}$$

The mean velocity of a gas is concerned in gas phenomena such as diffusion (effusion), viscosity and conduction of heat (p. 65). The root-mean-square velocity determines the magnitudes of the pressure of a gas and its specific heat capacity (p. 103).

From the distribution curve in fig. 3.10, it should be noted that the velocities of many molecules are higher than the most probable or mean or root-mean-square values. The r.m.s. velocity of the next-but-one lightest gas, helium, is just under 1 km s^{-1} at 15 °C, and molecules which have velocities greater than this will exist higher in the earth's atmosphere. A small percentage of these molecules will have velocities greater than about 11 km s^{-1}, the escape velocity from the earth's gravitational attraction, and are liable to escape from the earth's atmosphere (p. 60).

When the temperature of a given volume of gas is raised to a higher value T_2, the peak of the curve diminishes and the curve itself shifts to the right (fig. 3.10). The area between the curve and the axis of c is the same at T_1 and T_2, as it represents the total number of molecules. Although the velocities of the molecules are generally higher at T_2, the flatter peak of the curve shows that more molecules than before now have velocities closer to the most probable velocity. The velocity distribution thus tends to be more uniform as the temperature increases.

Mean free path

The *mean free path*, λ, of a molecule in a gas is defined as the average distance it travels between collisions. An approximate value of λ can be found by supposing σ is the effective diameter of a molecule moving in a constant direction with a velocity c.

In one second the molecule moves a distance c; during that time it makes collisions with all molecules in a cylinder of radius σ and axis that of the direction of motion. The volume of the cylinder is $\pi\sigma^2 c$; and if n is the number of molecules per unit volume, the number of collisions is $\pi\sigma^2 cn$.

$$\therefore \lambda = \frac{c}{\pi\sigma^2 cn} = \frac{1}{\pi\sigma^2 n} \quad . \quad . \quad . \quad . \quad . \quad . \quad . \quad \text{(i)}$$

More accurately, it has been shown that

$$\lambda = \frac{1}{\sqrt{2}\pi\sigma^2 n} \quad . \quad . \quad . \quad . \quad . \quad . \quad . \quad . \quad \text{(ii)}$$

From this relation, $\lambda \propto 1/n \propto 1/p$, where p is the pressure. Thus the mean free path increases as the pressure is lowered. At very low pressures, the molecules make few or no collisions with each other as they move about and strike the walls of the vessel. In this case the term "mean free path" has no meaning.

Viscosity of gas

The viscosity of a gas is explained on the kinetic theory by the continual transfer of momentum across layers of a gas while it is flowing along. A given layer of gas will receive faster molecules from a neighbouring layer above it, for example, and slower molecules from a layer below it. The momentum change of the layer is equivalent to a force on it, and this is the frictional or viscous force.

To find a formula for the coefficient of viscosity η on the kinetic theory, a gas is moving in a horizontal direction Oz. The number of molecules per unit volume moving along a direction Ox normal to Oz is $n/6$, where n is the number of molecules per unit volume. Suppose the average velocity of a molecule is $\bar{c}$. Then the number per second crossing an area A of a layer parallel to Oz is $nA\bar{c}/6$.

The molecules which enter a given layer in the gas are those which can be considered to come from a distance λ on either side, where λ is the mean free path. If the velocity gradient is dv/dx, the faster molecules from the layer above will bring a velocity of $v + \lambda dv/dx$ to the layer moving with velocity v, and the slower molecules from below a velocity of $v - \lambda dv/dx$.

$\therefore$ net change in momentum per second

$$= \frac{nA\bar{c}m}{6}\left[\left(v + \lambda\frac{dv}{dx}\right) - \left(v - \lambda\frac{dv}{dx}\right)\right]$$

$$\therefore \text{ frictional force } F = \tfrac{1}{3}nmA\bar{c}\lambda\frac{dv}{dx}$$

But
$$F = \eta A\frac{dv}{dx}$$

$$\therefore \eta = \tfrac{1}{3}nm\bar{c}\lambda = \tfrac{1}{3}\rho\bar{c}\lambda \quad \text{(i)}$$

where ρ is the density of the gas.

On p. 64 it was shown that $\lambda = 1/(\sqrt{2}\pi\sigma^2 n)$. With (i),

$$\eta = \tfrac{1}{3}\frac{m\bar{c}}{\sqrt{2}\pi\sigma^2}$$

The number of molecules n does not enter into this expression for η. Hence *η is independent of the pressure of the gas*. This surprising result is true for normal pressures, and is further experimental evidence for the kinetic theory of gases.

From (i), it follows, since $\bar{c}$ increases with temperature, that *η increases with temperature*. Maxwell's kinetic theory shows, in fact, that η should be proportional to $\sqrt{T}$, where T is the absolute temperature, but this relation is not borne out by experiment.

Thermal conductivity of gas

The thermal conductivity of a gas can be calculated on the kinetic theory by considering the transfer of energy (heat) across a given layer.

Following the treatment for the coefficient of viscosity, suppose $d\theta/dx$ is the constant temperature gradient and let us consider the molecules entering a given layer at a temperature θ. These will come from layers at a distance λ on either side, having respective temperatures of $\theta + \lambda d\theta/dx$ and $\theta - \lambda d\theta/dx$. The heat transfer per second dQ/dt will thus be given by, if c_V is the specific heat capacity of the gas at constant volume (see p. 101).

$$\frac{dQ}{dt} = \frac{nm\bar{c}A}{6}\left[\left(\theta + \lambda\frac{d\theta}{dx}\right) - \left(\theta - \lambda\frac{d\theta}{dx}\right)\right]c_V$$

$$\therefore \frac{dQ}{dt} = \tfrac{1}{3}nm\bar{c}A\lambda c_V\frac{d\theta}{dx}$$

But
$$\frac{dQ}{dt} = kA\frac{d\theta}{dx}$$

$$\therefore k = \tfrac{1}{3}nm\bar{c}\lambda c_V = \tfrac{1}{3}\rho\bar{c}\lambda c_V \quad \text{(i)}$$

From our previous results for viscosity, $\eta = \tfrac{1}{3}\rho\bar{c}\lambda$. Hence

$$k = \eta c_V \quad \text{(ii)}$$

Since η is independent of pressure, k should be independent of pressure. This is confirmed by experiment. More accurate theory has shown that $k = A\eta c_V$, where A is a constant depending on the atomicity of the gas.

Pumps

We conclude this chapter with a brief account of the application of the gas laws to pumps.

A bicycle pump is a form of *compression pump.* This forces air at atmospheric pressure into the inner tube of a tyre through a one-way valve. Suppose v is the volume in the barrel of the pump. Then after n strokes a volume nv *at atmospheric pressure* is forced into the tube. Thus if V is the volume of air at atmospheric pressure in the tube initially, the total volume of air at 1 atmosphere is $(V + nv)$. But this air now occupies a volume V at an increased pressure p say. Hence, from Boyle's law,

$$pV = 1 \times (V + nv)$$

$$\therefore p = 1 + \frac{nv}{V}$$

$$\therefore p - 1 = \frac{nv}{V}$$

Thus the increased pressure $(p - 1)$ is proportional to the number of strokes n.

Exhaust pump

Unlike the compression pump, an *exhaust pump* removes air from a vessel. During its action, two valves open and close alternately. Suppose a simple exhaust pump has a barrel of volume v and is connected to air initially at a pressure p inside a vessel of volume V. In the first "outward" stroke, the air expands from a volume V to a volume $(V + v)$. The pressure inside the vessel hence falls to a value p_1 given, from Boyle's law, by

$$p_1(V + v) = pV \quad \text{(i)}$$

On the "inward" stroke the second valve opens and the air in the barrel is then expelled.

On the second "outward" stroke, the pressure in the vessel falls from p_1 to a value p_2. From (i), p_2 is given by

$$p_2(V + v) = p_1 V$$

$$\therefore p_2 = p_1 \frac{V}{V + v} = p\left(\frac{V}{V + v}\right)^2$$

After n strokes, it can be seen that the pressure p_n in the vessel is given by

$$p_n = p\left(\frac{V}{V + v}\right)^n$$

There is a limit to the low pressure obtainable with this type of pump, because after a time the pressure in the vessel becomes too small to open the valve, and the pump then ceases to operate.

Rotary pump

Piston pumps have been supplanted by *rotary pumps,* which are simpler, faster, and produce higher vacua. The principle of the *Hyvac* type of rotary pump is illustrated in fig. 3.11. A cylindrical drum D is mounted eccentrically on a shaft S, which lies along the axis of a cylinder C, and the drum and cylinder are accurately

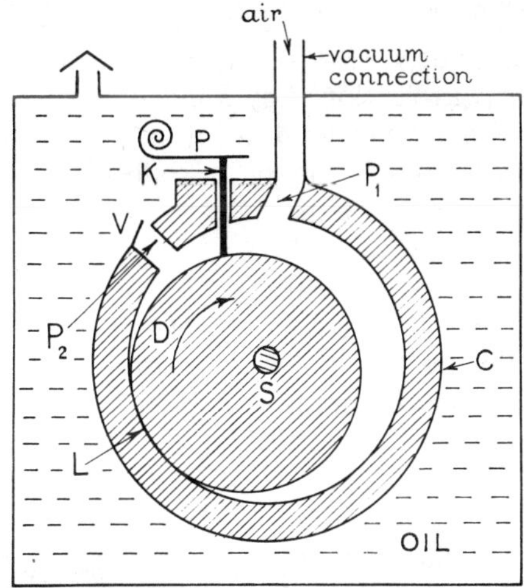

Fig. 3.11 Rotary pump

machined, so that the surface of the drum just touches the inner surface of the cylinder. Ports P_1, P_2 are cut in the cylinder to allow air to enter and leave it respectively, and the exit port P_2 is provided with a simple valve V. A scraping vane K is pressed on to the drum, between the ports, by a spring P, and the whole arrangement is immersed in oil of low vapour-pressure.

The drum D is rotated clockwise at a few hundred revolutions per minute. At the instant shown in fig. 3.11, the volume on the entrance side of the line of contact L is increasing, and air is therefore flowing through the vacuum connection into the cylinder. At the same time, the volume on the exit side of L is decreasing, and air is being driven out through the exit valve V. The scraper K prevents air flowing from the exit side of the pump to the entrance side. As D revolves further, the line of contact passes the exit port, which is then exposed to the vessel to be evacuated, and the atmospheric pressure closes the valve V. Shortly afterwards the line of contact passes the entrance port, and the volume on the entrance side of the drum becomes very small. It then increases, air is swept in through the vacuum connection, and is finally expelled through the exit valve V. The complete *Hyvac* pump comprises two of these units in tandem, in the same oil-filled case, and it can produce a pressure in a vessel as low as one-thousandth of a millimetre of mercury.

Measurement of low pressure by McLeod gauge

A pressure gauge widely used for measuring low pressures below a few mmHg (torr) was developed by McLeod. One form of McLeod gauge consists of a main tube T of glass connected to a mercury reservoir R, with a bulb B ending in a closed capillary tube C; a second capillary tube D, of the same bore as C, is sealed on to T (fig. 3.12). C and D are arranged to lie close together, and they are backed by a millimetre scale S.

To measure a low pressure p in a vessel, the latter is connected to the top of T. The reservoir R is now gently raised so that mercury flows into the bulb B. The air in B and C is then cut off, and when the mercury rises further, the air here is compressed.

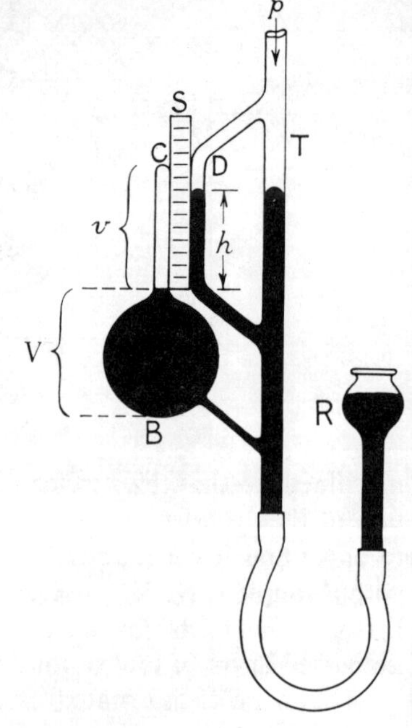

Fig. 3.12 Low-pressure measurement by McLeod gauge

Suppose P is the pressure of the air in the capillary tube C when the mercury completely fills B as shown. Then $P = p + h$, where h is the difference in levels of the mercury in C and D, and p is the pressure in the vessel connected to T. But since the mass of air in B and C originally had a volume $(V + v)$, where V is the volume of B and v the volume of C, *and the pressure was then p, the pressure in the vessel,* it follows from Boyle's law that

$$p(V + v) = Pv$$
$$\therefore p(V + v) = (p + h)v$$
$$\therefore pV = hv$$
$$\therefore p = \frac{v}{V}h$$

Thus *the pressure is proportional to h.* The scale S can hence be easily calibrated in terms of pressure.

Since $p = hv/V$, it can be seen that low pressures can be measured by making the volume V of the bulb B large and the volume v small.

The bores of C and D are made the same to eliminate surface tension forces between the mercury and the glass.

An alternative method of using the gauge is to bring the mercury in D level with the top of C, and then measure the difference in levels l of the mercury (fig. 3.13). The volume of the air trapped in C is now al, where a is the cross-section of C; and if its pressure is P, then $P \times al = p(V + v)$, applying Boyle's law, where p is the pressure to be measured. Now $P = p + l$.

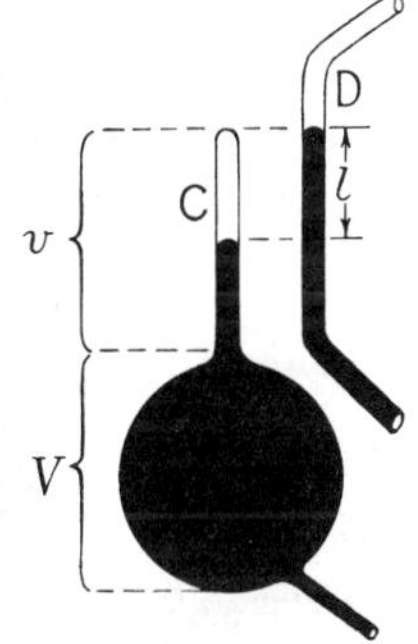

Fig. 3.13 McLeod gauge

$$\therefore (p + l)al = p(V + v)$$

$$\therefore p = \frac{al^2}{V + v - al}$$

Since v and al are very small compared with V, it follows that, to a very good approximation,

$$p = \frac{al^2}{V}$$

$$\therefore p \propto l^2$$

as a and V are constants. A wider range of pressures than before can hence be measured by bringing the mercury level to the top of C.

EXERCISE 3

Fill in the missing words in questions 1–5:

1. An ideal gas obeys Boyle's law and .... law.

2. For one mole of gas, the ratio pV/T is called the ....; its magnitude is about .... J mol^{-1} K^{-1}.

3. When a gas at a pressure p expands by a volume δV, the work done = ...; this is called .... work.

4. Gas pressure is calculated from kinetic theory by finding the .... of change of .... at the walls.

5. Gas pressure is proportional to the .... velocity of the molecules; the rate of diffusion (effusion) of a gas is proportional to the .... velocity of the molecules.

Gas laws

6. Define the *expansivity* (*volume coefficient*) of a gas. At constant pressure, a fixed mass of gas increases from a volume of 350 cm^3 at 0 °C to 360·25 cm^3 at 8 °C. Calculate the volume coefficient of the gas, and its volume at 100 °C.

7. A fixed mass of air has a pressure of 100 cm of mercury at 100 °C. Find its pressure (i) at 20 °C, (ii) at 0 °C, if the volume of the gas remains constant.

8. The volume of a fixed mass of gas at 50 °C is 80 cm^3. At what temperature does the volume become 100 cm^3, the pressure remaining constant?

9. What is the relation between the pressure, volume, and temperature of a fixed mass of gas? A quantity of oxygen gas has a volume of 250 cm^3 at 15 °C and 774 mmHg pressure. Calculate its volume at s.t.p. (0 °C and 760 mmHg).

10. The density of air is 1·29 kg m^{-3} at s.t.p. Calculate the gas constant of air for a mass of (i) 1 kg, (ii) 10 kg. What mass of air has a volume of 850 cm^3 at 27 °C and 750 mmHg pressure?

11. A gas expands from 80 cm^3 to 200 cm^3 under a constant external pressure of one atmosphere, assumed as 10^5 N m^{-2}. Find the work done by the gas against this pressure.

12. State Boyle's law and Charles' law and show how they may be combined to give the equation of state of an ideal gas.

Two glass bulbs of equal volume are joined by a narrow tube and are filled with a gas at s.t.p. When one bulb is kept in melting ice and the other is placed in a hot bath, the new pressure is 877·6 mm mercury. Calculate the temperature of the bath. (*L.*)

13. Give *brief* accounts of experiments which illustrate the relationship between the volume of a fixed mass of gas and (*a*) the pressure it exerts at a fixed temperature, (*b*) the temperature on a Celsius mercury thermometer at a fixed pressure. State the two "laws" which summarize the results.

A gas cylinder contains 6400 g of oxygen at a pressure of 5 atmospheres. An exactly similar cylinder contains 4200 g of nitrogen at the same temperature. What is the pressure on the nitrogen? (Molar masses: oxygen = 32, nitrogen = 28; assume that each behaves as a perfect gas.) (*O. and C.*)

14. State *Boyle's law* and *Charles' law*, and show how they lead to the gas equation $PV = RT$. Describe an experiment you would perform to measure the thermal expansion coefficient of dry air.

What volume of liquid oxygen (density 1140 kg m^{-3}) may be made by liquefying completely the contents of a cylinder of gaseous oxygen containing 100 litres of oxygen at 120 atmospheres pressure and 20 °C? Assume that oxygen behaves as an ideal gas in this latter region of pressure and temperature.

[1 atmosphere = 1·01 × 10^5 newton metre^{-2}; gas constant = 8·31 J mol^{-1} K^{-1}; molar mass of oxygen = 32·0.] (*O. and C.*)

15. The formula $pv = mrT$ is often used to describe the relationship between the pressure p, volume v, and temperature T of a mass m of a gas, r being a constant. Referring in particular to the experimental evidence how do you justify (*a*) the use of this formula, (*b*) the usual method of calculating T from the temperature t of the gas on the centigrade (Celsius) scale?

Two vessels each of capacity 1·00 litre are connected by a tube of negligible volume. Together they contain 0·342 g of helium at a pressure of 800 mm of mercury and temperature 27 °C. Calculate (i) a value for the constant r for helium, (ii) the pressure developed in the apparatus if one vessel is cooled to 0 °C and the other heated to 100 °C, assuming that the capacity of each vessel is unchanged. (*N.*)

16. State Boyle's law.

Describe how you would verify the law for dry air over a range of pressures from 0·5 to 1·5 atmospheres. Would the form of the apparatus you describe be suitable if the working range of pressure was 0·5 to 10 atmospheres? Give reasons for your answer.

Two glass vessels of equal volume are joined by a tube, the volume of which may be neglected. The whole is sealed and contains air at s.t.p. If one vessel is placed in boiling water at 100 °C and the other is placed in melting ice, what will be the resultant pressure of the air? (*N.*)

Kinetic theory

17. Show how the elementary kinetic theory of gases accounts for Boyle's law. Calculate the root mean square velocity of the molecules of hydrogen at 0 °C and at 100 °C. [Density of hydrogen at 0 °C and 760 mm mercury pressure = 0·09 kg m^{-3}.] (*L.*)

18. State the basic postulates of the kinetic theory of gases.

It can be shown that the pressure p exerted by a gas on the walls of its container is equal to $\frac{1}{3}\rho c^2$ where ρ is the density, and c is the root mean square velocity of the molecules. Explain why the root mean square velocity occurs in this expression. How is the concept of temperature introduced into the kinetic theory?

Show that the kinetic theory is consistent with any *two* of the following: (*a*) Graham's law of diffusion; (*b*) Dalton's law of partial pressures; (*c*) Avogadro's hypothesis.

Suggest an explanation of the following observation: A mixture of hydrogen and nitrogen is kept under pressure in a cylinder which has a leaky valve. Relatively more hydrogen escapes than nitrogen. (*O. and C.*)

19. Derive an expression connecting the pressure of a gas with its density and the mean square of the velocity of its molecules. State the assumptions that are required in the derivation of this expression.

Show that this expression is consistent with the gas equation pv/T = constant, provided a certain assumption is made. State this assumption.

Find the total energy per unit volume in a monatomic gas at standard temperature and pressure. Deduce an expression for the variation of this kinetic energy with temperature if the pressure remained constant. (Density of mercury = 13 600 kg m^{-3}.) (*C.*)

20. Explain what is meant by the *root mean square velocity* of the molecules of a gas. Use the concepts of the elementary kinetic theory of gases to derive an expression for the root mean square velocity of the molecules in terms of the pressure and density of the gas.

Assuming the density of nitrogen at s.t.p. to be 1·251 kg m^{-3}, find the root mean square velocity of nitrogen molecules at 127 °C. (*L.*)

21. State the assumptions that are made in the kinetic theory of gases and derive an expression for the pressure exerted by a gas which conforms to these assumptions, in terms of its density (ρ) and the mean square velocity (c^2) of its molecules.

Show (*a*) how *temperature* may be interpreted in terms of the theory, (*b*) how the theory accounts for Dalton's law of partial pressures. (*L.*)

22. Calculate the pressure in mm of mercury exerted by hydrogen gas if the number of molecules per cm^3 is $6{\cdot}80 \times 10^{15}$ and the root mean square speed of the molecules is $1{\cdot}90 \times 10^3$ m s^{-1}. Comment on the effect of a pressure of this magnitude (*a*) above the mercury in a barometer tube; (*b*) in a cathode ray tube. (Avogadro constant = $6{\cdot}02 \times 10^{23}$. Molecular weight of hydrogen = 2·02.) (*N.*)

23. Use a simple treatment of the kinetic theory of gases, stating any assumptions you make, to derive an expression for the pressure exerted by a gas on the walls of its container. Thence deduce a value for the root mean square speed of thermal agitation of the molecules of helium in a vessel at 0 °C. (Density of helium at s.t.p. = 0·1785 kg m^{-3}; 1 atmosphere = $1{\cdot}013 \times 10^5$ N m^{-2}.)

If the total translational kinetic energy of all the molecules of helium in the vessel is 5×10^{-6} joule, what is the temperature in another vessel which contains twice the mass of helium and in which the total kinetic energy is 10^{-5} joule? (Assume that helium behaves as a perfect gas.) (*O. and C.*)

24. Suggest reasons why the molecules in a gas at constant temperature are not all moving with the same speed.

A vessel of volume 500 cm^3 contains hydrogen at a pressure of 10^{-3} mmHg and at a temperature of 17 °C. Estimate (*a*) the number of molecules in the vessel, (*b*) their distance apart, on the average, (*c*) their root-mean-square speed, (*d*) the number of impacts they make per second on an area of 1 cm^2 of the wall of the vessel. ($R = 8{\cdot}3$ J mol^{-1} K^{-1}; number of molecules in a mole $= 6{\cdot}0 \times 10^{23}$; molecular weight of hydrogen $= 2{\cdot}0$; density of mercury $= 13\,600$ kg m^{-3}.) (*C.*)

4

Thermal Expansion of Solids and Liquids

SOLIDS

Observation shows that the length of a solid varies with temperature, and that a very large force is exerted by the solid when the expansion, or contraction, is opposed. The variation in length is temperature-compensated in designing balance wheels of watches and allowed for in the construction of railway lines or bridges. On the other hand, the expansion of metals with increasing temperature is utilized commercially in designing thermostats.

Molecular explanation of expansion

On molecular theory, we can explain the expansion of a solid with temperature rise by considering the variation of the intermolecular potential energy V with the distance r between two molecules, discussed on p. 55.

The lowest point X of the curve corresponds to the equilibrium separation r_0 of two molecules in the absence of thermal motion (fig. 4.1). At a higher temperature, T say, a vibrating molecule in a solid has an average potential energy of $\frac{1}{2}\,kT$ and an average kinetic energy of $\frac{1}{2}\,kT$, a total energy of kT (p. 59). Since the kinetic energy is zero at the extreme ends of the oscillation, the potential energy at these points is kT. The ends of the oscillations thus correspond to the points L and M respectively, shown in fig. 4.1. Now the curve is asymmetrical. The mean separation of the molecules is thus to the right of the vertical through X and roughly half-way between L and M. When the temperature rises further to a level of energy indicated by

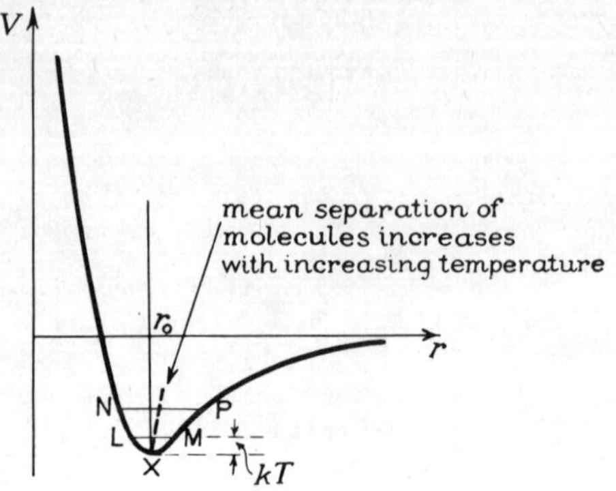

Fig. 4.1 Molecular explanation of expansion on heating

NP, the mean separation moves further to the right of the vertical through X, as shown roughly by the broken line. Thus *expansion* occurs. In a solid the intermolecular forces are very strong. This is why a large force is exerted when linear expansion or contraction is resisted (p. 78).

Linear expansivity

Different solids expand by different amounts when heated through the same temperature range. The *linear expansivity* (α) of a solid is defined as the *increase in length per unit length per °C temperature rise.* The magnitude of α depends on the initial temperature of the solid, and in standard practice this is 0 °C. For a solid, the variation of linear expansivity with an initial temperature other than 0 °C is extremely small.

When the solid is heated over a wide range of temperature, e.g. from 15 °C to 100 °C, we refer to the mean linear expansivity (α_m) in this temperature range. This may be defined as "the average increase in length per unit length per °C temperature change, in the temperature range 15 °C to 100 °C". Thus if the length of a solid increases from 120·0 cm at 15 °C to 120·2 cm at 100 °C, the mean coefficient in this temperature range

$$= \frac{120{\cdot}2 - 120{\cdot}0}{120 \times (100 - 15)} = 0{\cdot}000\,02 \text{ per °C} = 2 \times 10^{-5}\,°\text{C}^{-1}$$

Experiment shows that the mean linear expansivity of a solid varies slightly over different temperature ranges.

Unit and magnitude of linear expansivity

From the above calculation for the mean expansivity α_m it can be seen that

$$\alpha_m = \frac{\text{change in length}}{\text{original length} \times \text{change in temperature}}$$

The same formula holds for the linear expansivity α, the original length being that at 0 °C. If small changes in length are made at some temperature t °C, the linear expansivity at t can be expressed by the relation

$$\alpha = \frac{1}{l}\frac{dl}{dt}$$

where l is the length at t °C and dl/dt represents the change of l with temperature.

Since a "length" is present in the numerator and denominator of the expression for linear expansivity, their ratio is a number. Thus the *unit* of α is simply "per °C" or "$°C^{-1}$". Since a change of 1 °C = 1K (p. 3), *the SI unit of linear expansivity is "K^{-1}"*. The mean linear expansivity of brass is about 0·000018 K^{-1} or 18×10^{-6} K^{-1}. For iron it is about 11×10^{-6} K^{-1} and for glass about 8×10^{-6} K^{-1}. *Invar* is an alloy made from 64% iron, 36% nickel which has an extremely low linear expansivity of less than 1×10^{-6} at ordinary temperatures. It is therefore used in bimetal strips for thermostats.

Quartz and *fused silica* have extremely low linear expansivities, so that vessels of these materials have a negligibly small expansion when the temperature rises. *Pyrex*, the material used for cooking dishes, has a very small coefficient, which prevents the glass from cracking when the temperature rises or falls.

Formulae

Suppose a piece of brass 120 cm long at 0 °C is heated to a temperature of 40 °C. If the mean linear expansivity of brass is 18×10^{-6} K^{-1}, it follows by simple proportion that the expansion is $18 \times 10^{-6} \times 120 \times 40$ cm. In the same way, if a solid has a length l_0 at 0 °C, a mean linear coefficient α, and the temperature rises to t °C, then

$$\textit{increase in length} = \alpha l_0 t \qquad (1)$$

The new length l_t of the solid at t °C = original length + increase $= l_0 + \alpha l_0 t$.

$$\therefore l_t = l_0(1 + \alpha t) \qquad (2)$$

Further, since $(l_t - l_0)$ is the increase in length when a length l_0 at 0 °C increases to a length l_t at t °C, the mean linear coefficient is given by

$$\alpha = \frac{l_t - l_0}{l_0 \times t} \quad . \quad . \quad . \quad . \quad . \quad . \quad . \quad . \quad (3)$$

Similar formulae are obtained if we start with a length of the same solid at a temperature other than 0 °C; and since the mean coefficient of the solid has very nearly the same magnitude as before, we shall assume that the mean coefficient is still given by α. Thus if l_1 is a length of the solid at t_1 °C, and l_2 is the new length at a higher temperature t_2 °C,

$$\alpha = \frac{l_2 - l_1}{l_1 \times (t_2 - t_1)}$$

from which $$l_2 = l_1[1 + \alpha(t_2 - t_1)] \quad . \quad . \quad . \quad . \quad . \quad (4)$$

Over a small temperature range, the variation of length l of a solid with temperature t may be expressed by a linear relation such as (2) or (4). Over a wide temperature range, a square or cubic law such as $l_t = l_0(1 + at + bt^2 + ct^3)$ is followed, where $a = \alpha$, the linear expansivity defined in (3).

Laboratory method for linear expansivity

The mean linear expansivity of a metal may be measured in a school laboratory with apparatus similar in principle to that shown in fig. 4.2. One end A of a long metal rod is fixed against a stop S, while the other end B is free to expand.

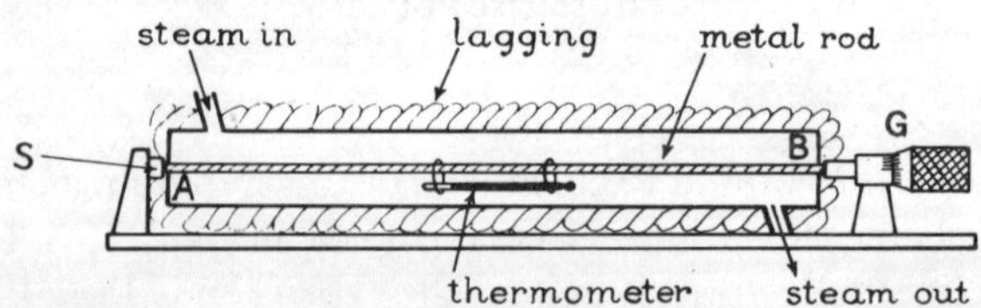

Fig. 4.2 Laboratory method for linear expansivity

Initially, the reading on a micrometer gauge G is observed when contact is just made with B. G is then turned back to allow expansion, and steam is passed for some time until the new reading on G is constant when contact is made with B again. The temperature is now observed.

If the expansion of the rod is 0·170 cm, the initial temperature of the rod is 15 °C and the final temperature 90 °C, and the original length of rod is 99·8 cm, then

$$\text{mean linear expansivity} = \frac{0{\cdot}170}{99{\cdot}8 \times (90 - 15)} = 23 \times 10^{-6}\ \text{K}^{-1}$$

Since the linear expansion of a metal is very small, a small error in measuring it would make a considerable percentage error in the final result for the expansivity. A form of micrometer gauge, which can measure small lengths very accurately, is therefore necessary. On the other hand, the original length of the bar is about 1 metre or 1000 mm long. An error of 1 or 2 mm in measuring the length would hence make only a very small percentage error in the final answer. Thus a metre rule may be used to measure the original length of the rod. Errors, and their influence on the final result of an experiment, are discussed more fully on p. 214.

Comparator method for measuring linear expansivity

An accurate method of measuring the linear expansivity of a solid is illustrated by fig. 4.3. It is known as a "comparator method" for a reason explained later, and is used at the International Bureau of Weights and Measures.

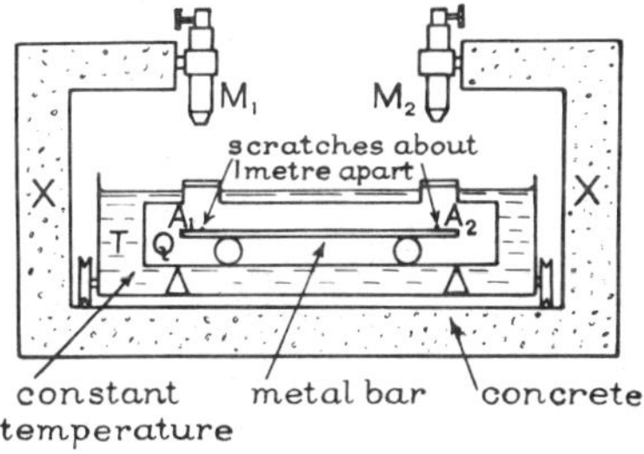

Fig. 4.3 Comparator method for linear expansivity

A standard metre bar P, one with marks on it exactly one metre apart, and the bar Q whose linear coefficient is required, are on separate trolleys inside a constant temperature enclosure T containing water at 0 °C. The bar Q has scratches on it about one metre apart. The trolleys move on wheels which enable the marks on P and Q to be brought in turn under the microscopes M_1 and M_2, which are supported by massive concrete pillars X.

The standard metre bar P is viewed first, and both microscopes are then adjusted by micrometer screws until the two marks A_1, A_2 on P are exactly under the cross-wires in each eyepiece. The distance between the cross-wires is then exactly one metre. P is now wheeled out of view, Q is brought into its place, and the two microscopes are adjusted until the two marks on Q are under the respective cross-wires. Knowing the total shift s of the two microscopes, the length l between the two marks on Q is obtained by adding (or subtracting) s from one metre. The temperature of the trough is now increased and maintained constant, and the microscopes are again adjusted so that the marks on Q are in line with the cross-wires. The net shift of the microscopes is equal to the expansion of a length l of the bar Q and, knowing the rise in temperature, the linear expansivity α can be calculated.

Owing to the massive concrete pillars on which they are mounted, the microscopes are not affected by any change in temperature. This method of determining the linear expansivity is known as the comparator method because the original length of the rod is obtained by comparing it with a standard metre bar.

Temperature change and metal scale

If a metal rule or scale is graduated correctly when the temperature is 10 °C, for example, the scale gives incorrect values when it is used at a different temperature, because of the change in length of the metal. The steel measuring tapes used by surveyors thus require a correction to be made to the recorded lengths after using the tapes.

Suppose that a metal tape is graduated correctly at 10 °C; 1 metre on the scale is then accurately 1 metre long at this temperature. At 20 °C, however, the 1 metre on the metal scale has expanded to $(1 + 10\alpha)$ metres, where α is the mean linear expansivity of the metal, since the temperature rise is 10 °C. Thus if a length l_1 measured by the metal tape at 20 °C is 40 m, the true length l_2 is greater than 40 m. Since $l_2 = l_1[1 + \alpha(t_2 - t_1)]$, the true length $= 40(1 + 10\alpha)$ metres. Using a steel tape, $\alpha = 11 \times 10^{-6}\ \mathrm{K}^{-1}$.

$$\therefore \text{true length} = 40(1 + 110 \times 10^{-6})\ \mathrm{m}$$

The length to be added to the observed length on the tape is hence $40 \times 110 \times 10^{-6}$ metre $= 0{\cdot}44$ cm.

Resistive force in heated bar

We have already mentioned the large force F set up in a metal rail or girder when its expansion or contraction is resisted. As we now see, the magnitude of F depends on the value of Young's modulus E for the metal concerned.

Suppose the metal is fixed between two immovable supports so that it cannot expand and has a length l. If its temperature rises to 50 °C say, it would normally expand by a length e, say, to a length l_1. The force F set up to resist expansion would be the force required to compress the bar by a length e from length l_1.

To calculate F, we know by definition that

$$E = \frac{\text{tensile stress}}{\text{tensile strain}} = \frac{F/A}{e/l_1}$$

where A is the cross-sectional area of the bar. Thus

$$F = EAe/l_1 = EAe/l$$

since l_1 is practically equal to l. But the increase in length e would be αlt (p. 75).

$$\therefore F = EA\alpha t$$

Suppose the bar has a cross-sectional area of 6 cm^2 or 6×10^{-4} m^2. Then, using $t = 50$ °C, $\alpha = 11 \times 10^{-6}$ K^{-1} for steel and $E = 2 \times 10^{11}$ N m^{-2},

$$F = 2 \times 10^{11} \times 6 \times 10^{-4} \times 11 \times 10^{-6} \times 50 \text{ N}$$

$$= 66 \times 10^3 \text{ N} = 660 \text{ kgf (approx)}$$

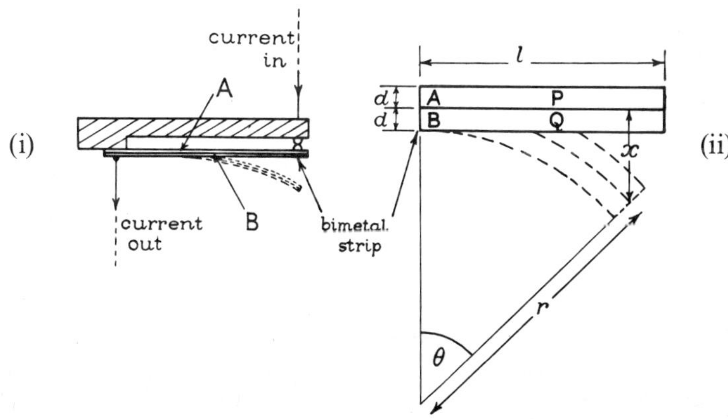

Fig. 4.4 Bimetal strip and deflection (exaggerated)

Bimetal strip

If equal lengths of two different metals A and B are welded together, they are said to form a *bimetal strip*. Figure 4.4 (i) illustrates the principle of a thermostat which utilizes a bimetal strip, as in an electric iron or an electric blanket. If A has a greater linear expansivity than B, and the temperature rise is excessive, the strip curls with A on the outside, since it expands more than B. The contacts are then broken, as indicated, and the current is cut off. Iron may be used as the metal for A and Invar for B.

The deflection x of a bimetal strip may be found from the length l and thickness d, assumed the same for both strips, their respective linear expansivities α_1, α_2, and the temperature rise t °C. If the length of the middle lines P and Q are taken as the expanded lengths of A and B respectively, then from the exaggerated fig. 4.4 (ii), if r is the radius of curvature,

$$\text{difference in expansion} = (r + \tfrac{1}{2}d)\theta - (r - \tfrac{1}{2}d)\theta = d\theta$$

But

$$\text{difference in expansion} = (\alpha_1 - \alpha_2)lt$$

$$\therefore \theta = \frac{(\alpha_1 - \alpha_2)lt}{d} = \frac{l}{r} \qquad (1)$$

$$\therefore r = \frac{d}{(\alpha_1 - \alpha_2)t} \qquad (2)$$

The depression x is related to l and r, assuming d negligible compared to x, by $x \,.\, 2r = l^2$, from the geometry of the circle. From (2),

$$\therefore x = \frac{l^2}{2r} = \frac{(\alpha_1 - \alpha_2)l^2 t}{2d}$$

Superficial (area) expansion

Besides expanding in length, solids expand in area when their temperature increases. Thus a metal cistern increases in surface area when the water it contains becomes hotter.

Suppose that a metal square of side l_0 at 0 °C is heated to a temperature of t °C. The new length of the side of the square then becomes $l_0(1 + \alpha t)$, where α is the linear expansivity, and hence the increase in the area of the square $= {l_0}^2(1 + \alpha t)^2 - {l_0}^2$.

$$\therefore \text{increase in area} = {l_0}^2(2\alpha t + \alpha^2 t^2)$$

The superficial expansivity β is defined as the "increase in area per unit area at 0 °C per °C temperature rise".

$$\therefore \beta = \frac{{l_0}^2(2\alpha t + \alpha^2 t^2)}{{l_0}^2 \times t} = 2\alpha + \alpha^2 t$$

But $\alpha^2 t$ is negligible compared to 2α, since α is of the order 10^{-5}.

$$\therefore \text{superficial expansivity} = 2 \times \text{linear expansivity}$$

Cubic (volume) expansion

When a solid, or a glass vessel, becomes hot, the volume changes as well as the length. Suppose that a *hollow cube* has each side of length l_0 at 0 °C and is heated to a temperature of t °C. If α is the linear expansivity, each side expands to $l_0(1 + \alpha t)$. The new volume of the cube is thus

$$l_0^3(1 + \alpha t)^3 = l_0^3(1 + 3\alpha t + 3\alpha^2 t^2 + \alpha^3 t^3)$$

and subtracting the original volume l_0^3,

$$\text{increase in volume} = l_0^3(3\alpha t + 3\alpha^2 t^2 + \alpha^3 t^3)$$

The coefficient of cubical (volume) expansion γ of a solid is defined as the "increase in volume of unit volume at 0 °C, when the temperature increases by 1 °C".

$$\therefore \gamma = l_0^3(3\alpha t + 3\alpha^2 t^2 + \alpha^3 t^3)/(l_0^3 \times t) = 3\alpha + 3\alpha^2 t + \alpha^3 t^2$$

But $3\alpha^2 t + \alpha^3 t^2$ is very small compared to 3α, since α is very small,

$$\therefore \gamma = 3\alpha$$

Thus the cubical expansivity of glass is (3 × 0·000 008) per °C if the linear expansivity of glass is 0·000 008 per °C.

Consider the expansion of a *solid* cube of side l_0 at 0 °C when it is heated to a temperature of t °C. The new volume is then $l_0^3(1 + \alpha t)^3$, since each side increases in length to $l_0(1 + \alpha t)$, and it can therefore be seen from above that *the expansion of a hollow vessel is exactly the same as if it were a solid vessel occupying the same volume.* We shall have occasion to use this result when we deal with the expansion of liquids (p. 82).

Examples

1. Explain what is meant by saying that the linear expansivity of brass is 0·000 019 per °C. How would you determine this coefficient over the range 0° – 100 °C? The barometric height, as read at 12 °C by a brass scale correct at 0 °C, is 765·2 mm. What is the actual height of the mercury column at 12 °C? (*N.*)

The length of 76·52 cm on the brass scale is correct only at 0 °C. The brass scale expands when the temperature rises to 12 °C, and the new length l_2 of 765·2 mm of brass at 0 °C is given by

$$\begin{aligned} l_2 &= l_1[1 + \alpha(t_2 - t_1)] \\ &= 765{\cdot}2[1 + 0{\cdot}000\,019(12 - 0)] \\ &= 765{\cdot}2 \times 1{\cdot}000\,228 = 765{\cdot}4 \end{aligned}$$

Thus the actual height of the mercury column at 12 °C is 765·4 mm.

2. Define Young's modulus for the material of a wire. The diagram shows an iron wire AB stretched inside a rigid brass framework and rigidly attached to it at both ends, A and B. The length of AB at 0 °C is 3 m and the diameter of the wire is 0·6 mm. What extra tension will be set up in the stretched wire when the temperature of the system is raised to 40 °C? (Take the linear expansivity of iron as $12 \times 10^{-6}\ K^{-1}$, the linear expansivity of brass as $18 \times 10^{-6}\ K^{-1}$; Young's modulus for iron as $2{\cdot}1 \times 10^{11}\ N\,m^{-2}$ (*L.*)

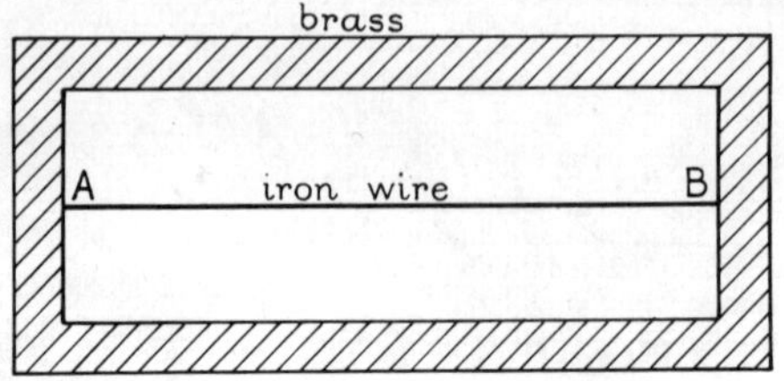

Fig. 4.5 Example

When the system is heated, the brass expands to a length $3(1 + 40\alpha) = 3(1 + 40 \times 18 \times 10^{-6}) = 3{\cdot}00216\ m$. The iron wire would normally have expanded to $3(1 + 40 \times 12 \times 10^{-6}) = 3{\cdot}00144\ m$.

$$\therefore \text{extension } e \text{ due to brass attachment} = 3{\cdot}00216 - 3{\cdot}00144 = 72 \times 10^{-5}\ m$$

With the usual notation, extra tension F in wire $= EAe/l$ (p. 79), where $A = \pi r^2 = \pi \times (3 \times 10^{-4})^2\ m^2$.

$$\therefore F = 2{\cdot}1 \times 10^{11} \times \pi \times (3 \times 10^{-4})^2 \times 72 \times 10^{-5}/3 \text{ newtons}$$

$$= 14\ N$$

LIQUIDS

Generally, liquids expand much more than solids when they are heated through the same temperature range (see below). The change in volume of a liquid with temperature is utilized in the domestic hot-water convection system. Fishes and other animal life are able to live in lakes when the temperature is below freezing-point on account of the special way in which the volume of water changes with temperature (p. 94).

The molecules of a liquid move about randomly inside the liquid, constantly exchanging neighbours. Unlike the case of a solid, where the molecules have only vibrational energy, the molecules in a liquid thus have translational energy, in addition to vibrational and rotational energy. The molecules in a gas have similar energies to those in a liquid, but the latter has a "more ordered state" as the inter-molecular attraction is much greater.

Cubic expansivity of liquid

A liquid occupies a *volume*. We thus deal with the volume or cubic changes of a liquid with temperature change.

The *cubic expansivity* γ of a liquid is defined as *the increase in volume per unit volume at 0 °C per °C temperature rise.* The magnitude of γ is about ten times the magnitude of the linear expansivity α of solids; for example, γ for water is about 2×10^{-4} K^{-1} at ordinary temperatures, whereas α for brass is about 2×10^{-5} K^{-1}.

The *mean* cubic expansivity γ_m of a liquid is the "average increase in volume per unit volume per °C temperature rise in the temperature range concerned". Thus if the volume of a given mass of water at 14 °C is 270 cm^3, and the volume increases to 273 cm^3 at 60 °C, the mean expansivity between 14 °C and 60 °C is given by

$$\gamma_m = \frac{273 - 270}{270 \times (60 - 14)} = 0{\cdot}000\,24\ \text{K}^{-1}$$

The table below shows how the mean cubic expansivity of water varies with different temperature ranges:

range	5 – 10 °C	10 – 20 °C	20 – 40 °C	40 – 60 °C	60 – 80 °C
γ_m	$0{\cdot}5 \times 10^{-4}$	1·5	3·0	4·6	5·9

Formulae

Suppose V_0 is the volume of a liquid at 0 °C and V_t is the volume at t°C. The mean cubic expansivity γ of the liquid between 0° and t°C is then given by

$$\gamma = \frac{V_t - V_0}{V_0 \times t} \qquad . \quad . \quad . \quad . \quad . \quad . \quad (1)$$

by definition of γ. Further, the increase in volume of the liquid from 0 °C, which is $(V_t - V_0)$, is given by

$$\text{increase in volume} = \gamma V_0 t \qquad . \quad . \quad . \quad . \quad (2)$$

and the volume V_t is thus $(V_0 + \gamma V_0 t)$.

$$\therefore V_t = V_0(1 + \gamma t) \qquad . \quad . \quad . \quad . \quad . \quad . \quad (3)$$

Similar formulae are obtained when the liquid is originally at a temperature other than 0 °C. For example, if V_1 is the volume of a liquid at t_1 °C, V_2 is the volume at t_2 °C, and γ is the mean coefficient of expansion in this temperature range, then $V_2 = V_1[1 + \gamma(t_2 - t_1)]$.

Absolute (true) and apparent cubic expansivities

A liquid must always be contained in a vessel of some kind. If the liquid is heated the vessel expands as well as the liquid, and the observed or *apparent* increase in volume of the liquid is equal to the difference between the actual or *absolute* expansion of the liquid and that of the vessel. If the vessel expands more than the liquid, the latter will appear to decrease in volume, and this can be observed to occur when a liquid is first warmed in a glass vessel, as the heat has then not yet reached the liquid. We must therefore distinguish between the absolute or true expansivity of a liquid and its apparent expansivity.

The *absolute* (*true*) *cubic expansivity* of a liquid γ is defined as the actual or true increase in volume per unit volume per °C temperature rise (p. 83). The *apparent expansivity* of a liquid γ_{app} is defined as the observed increase in volume per unit volume per °C temperature rise. The apparent expansivity thus takes into account the expansivity of the vessel containing the liquid. γ_{app} is hence less than γ, the absolute expansivity, and it can be shown that, approximately,

$$\gamma = \gamma_{app} + g$$

where g is the cubic expansivity of the material of the container (see below). Thus the value of g must be known if γ_{app} is measured in an experiment and the absolute coefficient γ is required. Suppose, for example, that the linear expansivity of glass is $8 \times 10^{-6}\ K^{-1}$, then the cubic expansivity of glass, $g = 3 \times 8 \times 10^{-6} = 24 \times 10^{-6}\ K^{-1}$ (p. 81). If the apparent expansivity of a liquid in a glass vessel is found by experiment to be $4{\cdot}5 \times 10^{-4}\ K^{-1}$, the absolute coefficient $\gamma = 4{\cdot}5 \times 10^{-4} + 0{\cdot}24 \times 10^{-4} = 4{\cdot}7 \times 10^{-4}\ K^{-1}$, to two significant figures.

To show $\gamma = \gamma_{app} + g$ approximately

Consider a vessel filled with a liquid of volume V_1 at t_1 °C to a level A (fig. 4.6(i)), and suppose an etching or scratch is made at this

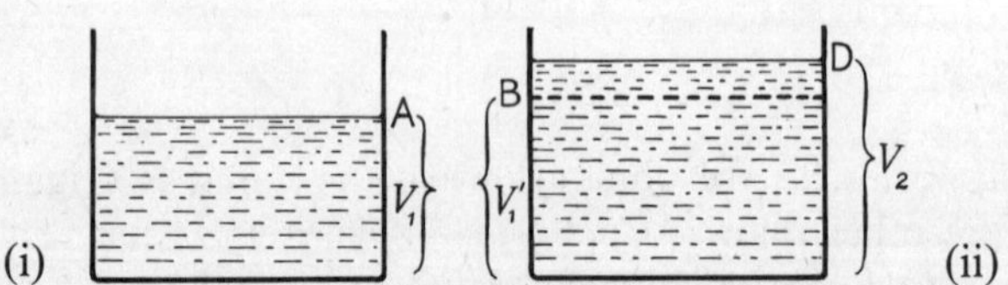

Fig. 4.6 Absolute and apparent expansivities

level on the container. If the liquid is heated to t_2 °C, the vessel expands to a volume V_1' at a level B, which is the new height of the etching if we assume the base of the vessel is fixed so that it cannot expand. At the same time the volume of the liquid increases to a value V_2 (fig. 4.6 (ii)). If the level of the liquid is now at D, the apparent expansion of the liquid is the volume between B and D, which is $(V_2 - V_1')$. The original volume of the liquid is the volume V_1. Hence the apparent expansivity of the liquid is given by

$$\gamma_{app} = \frac{\text{apparent increase in volume}}{\text{original volume} \times \text{temperature change}}$$

$$= \frac{V_2 - V_1'}{V_1(t_2 - t_1)}$$

where t_2, t_1 are the final and initial temperatures of the liquid. Now $V_2 = V_1[1 + \gamma(t_2 - t_1)]$ and $V_1' = V_1[1 + g(t_2 - t_1)]$, to a good degree of approximation.

$$\therefore \gamma_{app} = \frac{V_1[1 + \gamma(t_2 - t_1)] - V_1[1 + g(t_2 - t_1)]}{V_1(t_2 - t_1)}$$

$$= \frac{(\gamma - g)(t_2 - t_1)V_1}{V_1(t_2 - t_1)} = \gamma - g$$

$$\therefore \gamma = \gamma_{app} + g$$

Absolute expansivity and density of liquid

When a given mass of liquid is heated its volume increases and hence the density (mass/volume) of the liquid decreases. There is a simple relation between the density values of a liquid at various temperatures and its absolute cubic expansivity γ. Since this relation is the basis of many methods of determining γ, it should be noted carefully by the reader.

Let m = the mass of a liquid,

V_0, V_t = the volumes at 0 °C and t °C respectively,

and ρ_0, ρ_t = the densities at 0 °C and t °C respectively.

Since mass = volume × density, it follows that

$$m = V_0\rho_0 \quad \text{and} \quad m = V_t\rho_t$$

$$\therefore v_0\rho_0 = V_t\rho_t$$

$$\therefore \frac{\rho_0}{\rho_t} = \frac{V_t}{V_0}$$

But $V_t = V_0(1 + \gamma t)$, p. 83, and hence $\frac{V_t}{V_0} = 1 + \gamma t$.

$$\therefore \frac{\rho_0}{\rho_t} = 1 + \gamma t \quad . \quad . \quad . \quad . \quad . \quad . \quad (4)$$

When a liquid is heated from a temperature of t_1 °C to t_2 °C, it can be proved by an exactly similar method that

$$\frac{\rho_1}{\rho_2} = 1 + \gamma_m(t_2 - t_1)$$

where ρ_1, ρ_2 are the densities at t_1 °C, t_2 °C respectively, and γ_m is the mean coefficient of expansion of the liquid between these temperatures.

It should be noted that, in general, the density of a liquid is greater at a lower temperature than a higher one, since the volume of a given mass of liquid becomes smaller when the temperature decreases. Thus the density ρ_0 at 0°C is usually greater than the density ρ_t at t°C, where t is greater than zero.

Example

What do you understand by the statement "the mean cubic expansivity of mercury is 182×10^{-6} K^{-1}"? The density of mercury at 20 °C is 13 550 kg m^{-3}. At what higher temperature will this be in error by 1 per cent? (*N.*)

Suppose t °C is the higher temperature. Then density ρ_2 at this temperature

$$= 13\,550 - \frac{1}{100} \times 13\,550 = 13\,414{\cdot}5 \text{ kg m}^{-3}$$

But density at 20 °C, $\rho_1 = 13\,550$ kg m^{-3}

and $\frac{\rho_1}{\rho_2} = 1 + \gamma(t_2 - t_1)$, with the usual notation.

$$\therefore \frac{13\,550}{13\,414{\cdot}5} = \frac{100}{99} = 1 + 182 \times 10^{-6}(t - 20)$$

$$\therefore t - 20 = \frac{1}{99 \times 182 \times 10^{-6}} = 56$$

$$\therefore t = 76 \text{ °C}$$

Dulong and Petit balancing-column method

We now consider methods of measuring the cubic expansivities of liquids. In some methods, considered later, the absolute expansivity is found by measuring the apparent expansivity and then adding the cubic expansivity of the material of the container. In 1808, however, Dulong and Petit devised a method for determining the absolute expansivity of a liquid *without* any reference to the cubic expansivity of the container.

Figure 4.7 illustrates the principle of the method. The liquid is placed in a U-tube, one limb of which is maintained at a constant temperature t_1 °C, and the other at a higher temperature t_2 °C. When conditions are steady the liquid level Y in the latter limb will be higher than the level D of the liquid in the other limb.

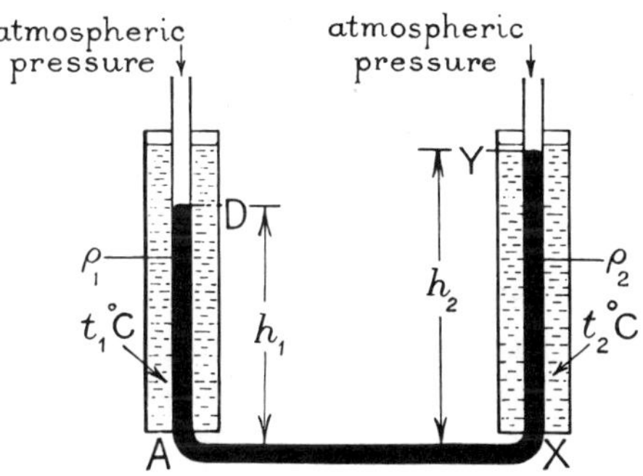

Fig. 4.7 Principle of balancing-column method—Dulong and Petit

Suppose that the base AX of the U-tube is horizontal, and that the height AD is h_1, and the height XY is h_2. Then, since A and X are points in the same horizontal level of the liquid,

$$\text{the pressure at A} = \text{the pressure at X} \quad . \quad . \quad . \quad \text{(i)}$$

Now the pressure p at a depth h below the surface of a liquid of density ρ is given by $p = h\rho g$, where g is the acceleration due to gravity. Thus, since both limbs of the U-tube are open, the pressure at A is $B + h_1\rho_1 g$, where B is the atmospheric pressure and ρ_1 is the density of the liquid at t_1 °C; and the pressure at X is $B + h_2\rho_2 g$, where ρ_2 is the density of the liquid at t_2 °C. Consequently, from (i),

$$B + h_1\rho_1 g = B + h_2\rho_2 g \quad . \quad . \quad . \quad . \quad \text{(ii)}$$

$$\therefore h_1\rho_1 = h_2\rho_2$$

$$\therefore \frac{\rho_1}{\rho_2} = \frac{h_2}{h_1}$$

But $$\frac{\rho_1}{\rho_2} = 1 + \gamma(t_2 - t_1)$$

where γ is the mean expansivity of the liquid.

$$\therefore 1 + \gamma(t_2 - t_1) = \frac{h_2}{h_1}$$

$$\therefore \gamma(t_2 - t_1) = \frac{h_2}{h_1} - 1 = \frac{h_2 - h_1}{h_1}$$

$$\therefore \gamma = \frac{h_2 - h_1}{h_1(t_2 - t_1)} \quad . \quad . \quad . \quad . \quad . \quad \text{(iii)}$$

Thus if $(h_2 - h_1)$, h_1, t_2, t_1 are measured, the *absolute* expansivity γ of the liquid can be calculated.

It will be noted that the cubic expansivity of the vessel does not enter into the calculation for γ, or in the final expression for γ given in (iii). Of course, the limb XY of the vessel expands if it is heated; but the pressure at X depends only on the *height* of the liquid above this point, and the level Y of the liquid accordingly adjusts itself to make the pressure at X equal to the pressure at A. Dulong and Petit's "balancing column" method, as it is sometimes called, thus enables a direct determination of the absolute cubic expansivity of a liquid to be made.

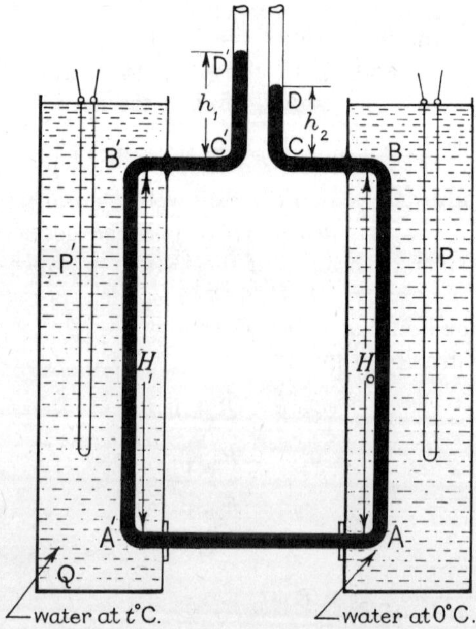

Fig. 4.8 Absolute cubic expansivity of mercury

Absolute expansivity of mercury—Callendar and Moss

From the theory of Dulong and Petit, equation (iii), the absolute expansivity $\gamma = (h_2 - h_1)/[h_1(t_2 - t_1)]$. Now $(h_2 - h_1)$ is the difference in levels of the liquid at D, Y (fig. 4.7), and is very small, perhaps 0·4 cm. Thus if γ is to be determined with accuracy, the difference in levels must be measured very accurately. Hence a travelling microscope must be used. Since h_1 is a comparatively big length, e.g. 65 cm, a small error in this measurement will make very little difference to the value of γ. Hence h_1 may be measured with a metre rule.

Callendar and Moss made a very accurate determination of the absolute expansivity of mercury by the method of balancing columns (fig. 4.8). Mercury was contained in a glass tube whose open ends were brought together at D′, D, and the column of mercury in A′B′ was maintained at a constant high temperature by a liquid bath Q. The mercury in AB was kept at the constant temperature 0 °C by continuously circulating ice-cold water round it. The rest of the mercury was also maintained at 0 °C by wrapping pieces of blotting-paper dipped in melting ice round A′A, B′C′, C′D′, CD.

Suppose A′B′ = H_1, AB = H_0; C′D′ = h_1, CD = h_2; the temperatures of A′B′, AB = t°C, 0°C; ρ_0, ρ_t = the respective densities of mercury at 0 °C and t °C; and B = the atmospheric pressure. Then *if* A′A *is kept horizontal*,

pressure at A′ = pressure at A

$$\therefore B + H_1\rho_t g + h_1\rho_0 g = B + H_0\rho_0 g + h_2\rho_0 g$$

$$\therefore H_1\rho_t = (H_0 + h_2 - h_1)\rho_0$$

$$\therefore \frac{\rho_0}{\rho_t} = \frac{H_1}{H_0 + h_2 - h_1}$$

But

$$\frac{\rho_0}{\rho_t} = 1 + \gamma t$$

$$\therefore 1 + \gamma t = \frac{H_1}{H_0 + h_2 - h_1}$$

$$\therefore \gamma = \frac{H_1 - H_0 + h_1 - h_2}{(H_0 + h_2 - h_1)t}$$

The heights H_1, H_0, and the length $(h_1 - h_2)$ were measured accurately by a cathetometer, which is a travelling microscope moving in a vertical direction. The temperatures were measured by resistance thermometers, P′, P. In order to increase the difference in levels of the mercury, thus increasing the sensitivity, Callendar and Moss used six "hot" and six "cold" tubes (instead of one "hot" and one "cold" tube as shown in fig. 4.8; they connected them so that six of the tubes were maintained at the same high temperature and six were at a temperature of 0 °C. Regnault had carried out a similar experiment to measure the absolute coefficient of expansion of mercury in 1847, but had used only one "hot" and one "cold" tube.

Converting barometric heights to 0 °C

The barometric height is recorded at meteorological stations at intervals during the day. Since mercury expands when its temperature increases, the temperature must also be recorded at the time the barometric height is measured, and for purposes of comparison, it is usual to calculate the height of mercury *at* 0 °C which would have the same pressure as that observed. The "correction" of the barometric height to 0 °C involves the linear expansivity of the steel scale provided with the barometer, as well as the cubic expansivity of the mercury, and the following example illustrates how the correction is made.

Example

If l is the linear expansivity of a solid, explain carefully why $3l$ may be used as the cubic expansivity. The height of a mercury barometer read with a steel scale is 754 mm at 20 °C. What will it read at 0 °C? (Linear expansivity of steel = $12 \times 10^{-6}\,\mathrm{K}^{-1}$; cubic expansivity of mercury = $182 \times 10^{-6}\,\mathrm{K}^{-1}$). (*O. and C.*)

Suppose that the steel scale reads correctly at 0 °C. Then at 20 °C the reading of 754 mm on the steel scale is smaller than the true length h_t which is given by

$$h_t = 754(1 + 12 \times 10^{-6} \times 20) \qquad \text{(i)}$$

Let h_0 = the required height of the barometer at 0 °C.
Then, since the pressures at 0 °C are the same,

$$h_0\rho_0 g = h_t\rho_t g$$

where ρ_0, ρ_t are the densities of mercury at 0 °C, 20 °C respectively.

$$\therefore h_0 = h_t\frac{\rho_t}{\rho_0} = \frac{h_t}{1 + 20\gamma}$$

since $\rho_t/\rho_0 = 1/(1 + \gamma t)$, where γ = cubic expansivity of mercury = $182 \times 10^{-6}\,\mathrm{K}^{-1}$.
But, from (i), $h_t = 754(1 + 12 \times 10^{-6} \times 20)$.

$$\therefore h_0 = \frac{754(1 + 12 \times 10^{-6} \times 20)}{1 + 182 \times 10^{-6} \times 20}$$

$$= 754(1 + 12 \times 10^{-6} \times 20)(1 + 182 \times 10^{-6} \times 20)^{-1}$$

$$= 754(1 + 12 \times 10^{-6} \times 20)(1 - 182 \times 10^{-6} \times 20), \text{ by binomial expansion,}$$

$$= 754[1 - (182 - 12) \times 10^{-6} \times 20], \text{ approx} = 751{\cdot}44 \text{ mm}$$

Generally, if α is the linear expansivity of the metal barometer scale and ρ is the density of mercury, then, if h_t is the observed barometer reading,

$$h_0 = h_t \frac{1 + \alpha t}{1 + \gamma t} = h_0[1 - (\gamma - \alpha)t]$$

to a good approximation.

Apparent expansivity by weight thermometer

One of the simplest methods of determining the *apparent* expansivity of a liquid is to fill a density bottle P full of liquid. Instead of a density bottle, a glass vessel Q of the shape shown in fig. 4.9 is sometimes used; Q is known as a "weight thermometer".

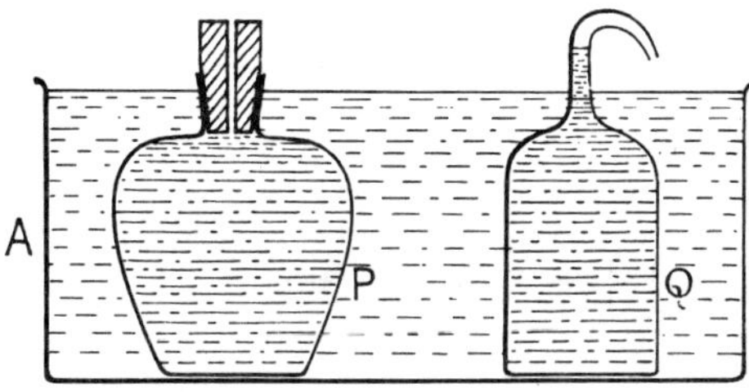

Fig. 4.9 Weight thermometers for apparent expansivity

The mass of liquid filling the vessel is first measured. The vessel is then placed in a water bath A and is heated to a known temperature. Since the liquid expands, some of it is expelled from the vessel through the opening at the top, and when the vessel is reweighed, the mass of liquid left in the vessel is determined. The apparent expansivity is then given by

$$\gamma_{\text{app}} = \frac{\textit{mass expelled}}{\textit{mass}\ \text{LEFT} \times \textit{temperature rise}} \quad . \quad . \quad . \quad . \quad \text{(i)}$$

To prove relation (i) above

Suppose m_1, m_2 are the masses of the liquid filling the bottle at temperatures t_1 °C, t_2 °C, where t_2 is greater than t_1; V_1, V_2 are the respective volumes of the liquid; and ρ_1, ρ_2 are the respective densities of the liquid. Then, since mass = volume × density,

$$m_1 = V_1\rho_1;\ m_2 = V_2\rho_2$$

$$\therefore \frac{m_1}{m_2} = \frac{V_1\rho_1}{V_2\rho_2} = \frac{V_1}{V_2} \times \frac{\rho_1}{\rho_2} \quad . \quad . \quad . \quad . \quad . \quad . \quad . \quad \text{(i)}$$

Now the ratio ρ_1/ρ_2 concerns the mean coefficient of expansion γ of the liquid between the temperatures t_1 °C, t_2 °C, and $\rho_1/\rho_2 = 1 + \gamma(t_2 - t_1)$, p. 87. The ratio V_1/V_2

concerns the *mean cubic expansivity g of the material of the vessel,* since a liquid always conforms to the shape of the vessel containing it. Thus

$$V_2 = V_1[1 + g(t_2 - t_1)]$$

and hence
$$\frac{V_1}{V_2} = \frac{1}{1 + g(t_2 - t_1)}$$

Substituting for V_1/V_2 and ρ_1/ρ_2 in (i), we have

$$\frac{m_1}{m_2} = \frac{1}{1 + g(t_2 - t_1)} \times [1 + \gamma(t_2 - t_1)] = \frac{1 + \gamma(t_2 - t_1)}{1 + g(t_2 - t_1)}$$

Cross-multiplying and simplifying for γ, we obtain

$$\gamma = \frac{m_1 - m_2}{m_2(t_2 - t_1)} + \frac{m_1}{m_2}g \qquad \text{(ii)}$$

A convenient approximation can be made in this expression for γ by considering the results of a typical experiment. The original mass of liquid m_1 may be 62·41 g, and the mass of liquid m_2 left after heating 59·70 g. Thus

$$\frac{m_1}{m_2}g = \frac{62{\cdot}41}{59{\cdot}70}g, \quad \text{and hence} \quad \frac{m_1}{m_2}g = g$$

to a good degree of approximation. Hence, from (ii),

$$\gamma = \frac{m_1 - m_2}{m_2(t_2 - t_1)} + g$$

$$\therefore \gamma - g = \frac{m_1 - m_2}{m_2(t_2 - t_1)}$$

But
$$\gamma - g = \gamma_{app}$$

where γ_{app} is the apparent expansivity of the liquid.

$$\therefore \gamma_{app} = \frac{m_1 - m_2}{m_2(t_2 - t_1)}$$

Since $(m_1 - m_2)$ is the "mass of liquid expelled" on heating, and m_2 is the "mass of liquid *left*", it follows that

$$\gamma_{app} = \frac{\textit{mass expelled}}{\textit{mass}\ \text{LEFT} \times \textit{temperature rise}}$$

The specific gravity bottle, or weight thermometer, thus enables the apparent expansivity of a liquid to be easily found. If the absolute expansivity of the liquid is required, the cubic expansivity g of the vessel's material must be added to the apparent expansivity; but this method does not give a very accurate value for the absolute expansivity, as g is not known to a high degree of accuracy. If the specific gravity bottle were made of quartz or Pyrex, the magnitude of g would then be so small as to be negligible in comparison with the magnitude of γ_{app}. In this case the absolute expansivity of the liquid can be accurately measured by the weight thermometer method.

Examples

1. When a glass weight thermometer filled with mercury at 0 °C is heated to 100 °C, 7·785 g of mercury overflow and 450·0 g remain in the thermometer. Using glycerine in place of mercury, the corresponding figures are 2·173 g and 41·00 g. The cubic expansivity of mercury is $183 \times 10^{-6}\,K^{-1}$. Determine (*a*) the linear expansivity of the glass, (*b*) the cubic expansivity of the glycerine. (*N.*)

(*a*) Apparent expansivity of mercury

$$\gamma_{app} = \frac{\text{mass expelled}}{\text{mass left} \times \text{temperature rise}}$$

$$= \frac{7{\cdot}785}{450{\cdot}0 \times 100} = 173 \times 10^{-6}\,K^{-1}$$

But the absolute expansivity of mercury $\gamma = \gamma_{app} + g$, where g is the cubic expansivity of glass.

$$\therefore g = \gamma - \gamma_{app} = (183 - 173) \times 10^{-6}$$

$$= 10 \times 10^{-6}\,K^{-1}$$

$\therefore$ linear expansivity of glass $= \frac{1}{3} \times g = \frac{1}{3} \times 10 \times 10^{-6} = 3{\cdot}3 \times 10^{-6}\,K^{-1}$

(*b*) Apparent expansivity γ_{app} of glycerine

$$= \frac{\text{mass expelled}}{\text{mass left} \times \text{temperature rise}}$$

$$= \frac{2{\cdot}173}{41{\cdot}00 \times 100} = 530 \times 10^{-6}\,K^{-1}$$

$\therefore$ absolute expansivity of glycerine $= \gamma_{app} + g$

$$= 530 \times 10^{-6} + 10 \times 10^{-6} = 540 \times 10^{-6} = 54 \times 10^{-5}\,K^{-1}$$

2. Distinguish between the real and apparent expansivity of a liquid. Deduce an expression relating the density of a liquid, at different temperatures, with its expansivity.

A glass bottle, volume 50 cm^3 at 0 °C, is filled with paraffin at 15 °C. What is the mass of the paraffin? The density of paraffin at 0 °C is 820 kg m^{-3}; the true expansivity of paraffin for the range 0° to 15 °C is $9 \times 10^{-4}\,K^{-1}$; and the linear expansivity of glass is $9 \times 10^{-6}\,K^{-1}$. (*N.*)

Cubic expansivity of glass $= 3 \times 9 \times 10^{-6} = 27 \times 10^{-6}\,K^{-1}$

$\therefore$ volume of bottle at 15 °C $= 50(1 + 27 \times 10^{-6} \times 15) \times 10^{-6}\,m^3$

Also, density of paraffin at 15 °C $= \dfrac{\rho_0}{1 + \gamma t} = \dfrac{820}{1 + 9 \times 10^{-4} \times 15}\,kg\,m^{-3}$

$\therefore$ mass of paraffin at 15 °C = volume × density

$$= 50(1 + 27 \times 10^{-6} \times 15) \times 10^{-6} \times \frac{820}{(1 + 9 \times 10^{-4} \times 15)}$$

$$= 50{\cdot}02 \times \frac{820}{1{\cdot}0135} \times 10^{-6}\,kg$$

$$= 40{\cdot}5 \times 10^{-3}\,kg = 40{\cdot}5\,g \text{ (approx)}$$

Anomalous expansion of water

Most liquids expand when they are heated, so that the density (mass/volume) decreases with increasing temperature. Water, however, is one of the few exceptions to this rule. It is known by experiment that the density of water at normal pressures *increases* from 0 °C to 4 °C, i.e. the density variation is anomalous in this temperature range. Beyond 4 °C, the density of water decreases with increasing temperature, like most liquids, so that water has a maximum density at 4 °C. Figure 4.10 illustrates roughly the variation of the density of water, and the volume of a given mass, with temperature.

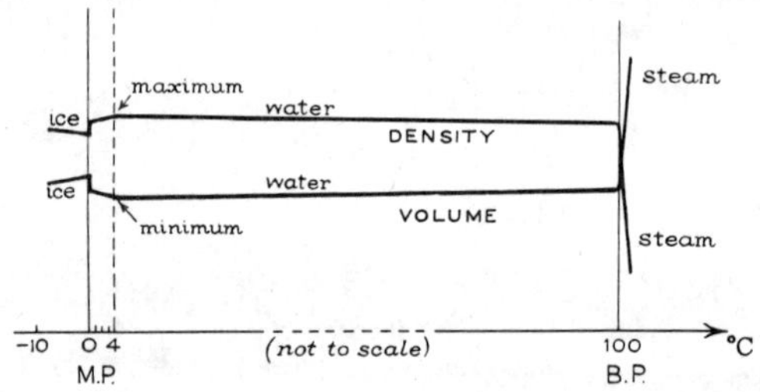

Fig. 4.10 Variation of density and volume of water

The temperature of maximum density of water was first obtained by Hope in 1804; he used a tall cylinder containing water, with a freezing mixture F round the middle (fig. 4.11). Observations were taken on two thermometers T_1, T_2, which were placed respectively in the upper and lower parts of the water, and it was found that the lower thermometer first fell from 15 °C, say, to a temperature of 4 °C,

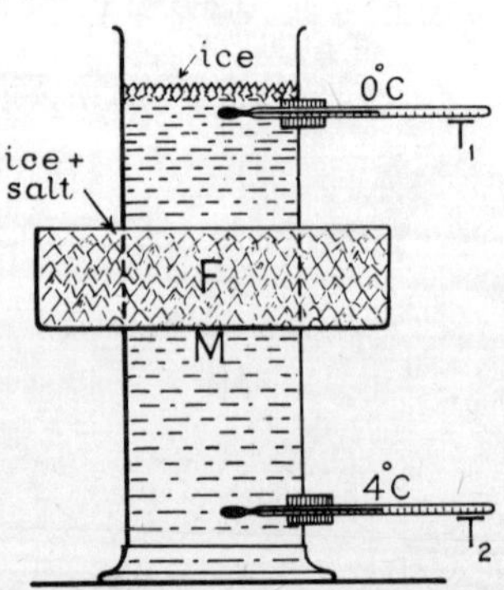

Fig. 4.11 Hope's experiment on anomalous expansion of water

and then remained constant at this temperature. After this the upper thermometer T_1 began to fall in temperature from 15 °C, until ice was formed at the top, when the temperature reached 0 °C.

The following table gives the densities of water at some temperatures, in units of kg m^{-3}:

0 °C	2 °C	4 °C	6 °C	80 °C	90 °C	100 °C
999·84	999·94	999·97	999·94	971·8	965·3	958·4

Density of ice at 0°C = 920 kg m^{-3} (approx).
Density of steam at 100°C = 0·60 kg m^{-3} (approx).

EXERCISE 4

Solids

1. Define the *linear expansivity* of a solid, and describe a method by which it may be measured.

Show how the superficial expansivity can be derived from this value.

A "thermal tap" used in certain apparatus consists of a silica rod which fits tightly inside an aluminium tube whose internal diameter is 8 mm at 0 °C. When the temperature is raised, the fit is no longer exact. Calculate what change in temperature is necessary to produce a channel whose cross-section is equal to that of a tube of 1 mm internal diameter. [Linear expansivity of silica = 8×10^{-6} K^{-1}. Linear expansivity of aluminium = 26×10^{-6} K^{-1}.] (*O. and C.*)

2. A steel wire 8 metres long and 4 mm in diameter is fixed to two rigid supports. Calculate the increase in tension when the temperature falls 10 °C. [Linear expansivity of steel = 12×10^{-6} K^{-1}, Young's modulus for steel = 2×10^{11} N m^{-2}.] (*O. and C.*)

3. Describe in detail how you would determine the linear expansivity of a metal rod or tube. Indicate the chief sources of error and discuss the accuracy you would expect to obtain.

A steel cylinder has an aluminium alloy piston and, at a temperature of 20 °C when the internal diameter of the cylinder is exactly 10 cm, there is an all-round clearance of 0·05 mm between the piston and the cylinder wall. At what temperature will the fit be perfect? [The linear expansivity of steel and the aluminium alloy are $1{\cdot}2 \times 10^{-5}$ and $1{\cdot}6 \times 10^{-5}$ K^{-1} respectively.] (*O. and C.*)

4. Describe an accurate method for determining the linear expansivity of a solid in the form of a rod. The pendulum of a clock is made of brass whose linear expansivity is $1{\cdot}9 \times 10^{-5}$ per °C. If the clock keeps correct time at 15 °C, how many seconds per day will it lose at 20 °C? (*O. and C.*)

5. Explain carefully what you understand by the term *temperature*. Describe how you would measure, to an accuracy of one hundredth of a centigrade degree, a steady temperature.

A bimetallic strip is composed of two metals A and B, each 0·2 mm thick, which are welded together all along their common surface. The linear expansivity of the metals per °C are $1{\cdot}8 \times 10^{-5}$ for A and 4×10^{-6} for B. The strip is straight and is 10 cm long when the temperature is 15 °C, and one of its ends is fixed. Estimate (*a*) the radius of curvature of the common surface when the temperature rises to 30 °C, (*b*) the corresponding deflection of the free end of the strip. (*O.*)

Liquids

6. Explain the statement: *the absolute expansivity of mercury is* $1{\cdot}81 \times 10^{-4}\,K^{-1}$. Describe an experiment to test the accuracy of this value. Why is a knowledge of it important?

Calculate the volume at 0 °C required in a thermometer to give a degree of length 0·15 cm on the stem, the diameter of the bore being 0·24 mm. What would be the volume of this mercury at 100 °C? [The linear expansivity of the glass may be taken as $8{\cdot}5 \times 10^{-6}\,K^{-1}$.] (*L.*)

7. Define the *cubic expansivity* of a liquid. Find an expression for the variation of the density of a liquid with temperature in terms of its expansivity.

Describe, without experimental details, how the cubic expansivity of a liquid may be determined by the use of balanced columns.

A certain Fortin barometer has its pointers, body and scales made from brass. When it is at 0 °C it records a barometric pressure of 760 mmHg. What will it read when its temperature is increased to 20 °C if the pressure of the atmosphere remains unchanged?

[Cubic expansivity coefficient of mercury = $1{\cdot}8 \times 10^{-4}\,K^{-1}$; linear expansivity of brass = $2 \times 10^{-5}\,K^{-1}$.] (*O. and C.*)

8. Define linear and cubic expansivity, and derive the relation between them for a particular substance. Describe, and give the theory of, a method for finding *directly* the absolute cubic expansivity of a liquid. The bulb of a mercury-in-glass thermometer has a volume of 0·5 cm^3 and the distance between progressive degree marks is 2 mm. If the linear expansivity of glass is $10^{-5}\,K^{-1}$, and the cubic expansivity of mercury is $1{\cdot}8 \times 10^{-4}\,K^{-1}$, find the cross-sectional area of the bore of the stem. (*C.*)

9. Describe an experiment to determine the absolute (real) expansivity of paraffin between 0 °C and 50 °C, using a weight thermometer made of glass of known cubic expansivity. Derive the formula used to calculate the result.

A glass capillary tube with a uniform bore contains a thread of mercury 100·0 cm long, the temperature being 0 °C. When the temperature of the tube and mercury is raised to 100 °C the thread is increased in length by 1·65 cm. If the mean coefficient of absolute expansion of mercury between 0 °C and 100 °C is 0·000182 K^{-1}, calculate a value for the linear expansivity of glass. (*N.*)

10. Describe and explain how the absolute expansivity of a liquid may be determined without a previous knowledge of any other expansivity.

Aniline is a liquid which does not mix with water, and when a small quantity of it is poured into a beaker of water at 20 °C it sinks to the bottom, the densities of the two liquids at 20 °C being 1021 and 998 kg m^{-3} respectively. To what temperature must the beaker and its contents be uniformly heated so that the aniline will form a globule which just floats in the water? (The mean absolute expansivity of aniline and water over the temperature range concerned are 0·00085 and 0·00045 K^{-1} respectively.) (*L.*)

11. Why is mercury used as a thermometric fluid? Compare the advantages and disadvantages of the use of a mercury-in-glass thermometer and a platinum resistance thermometer to determine the temperature of a liquid at about 300 °C.

A dilatometer having a glass bulb and a tube of uniform bore contains 150 g of mercury which extends into the tube at 0 °C. How far will the meniscus rise up the tube when the temperature is raised to 100 °C if the area of cross-section of the bore is 0·8 mm^2 at 0 °C? Assume that the density of mercury at 0 °C is 13·6 g cm^{-3}, that the expansivity of mercury is 1·82 × 10^{-4} K^{-1}, and that the linear expansivity of glass is 1·1 × 10^{-5} K^{-1}. *(N.)*

12. Describe how to measure the apparent expansivity of a liquid using a weight thermometer. Show how the result can be calculated from the observations.

A specific gravity bottle of volume 50·0 cm^3 at 0 °C is filled with glycerine at 20 °C. What mass of glycerine is contained in the bottle if the density of glycerine at 0 °C is 1260 kg m^{-3}, and its real expansivity is 5·2 × 10^{-4} K^{-1}? Assume that the linear expansivity of the glass is 8 × 10^{-6} K^{-1}. *(N.)*

13. Describe in detail how a reliable value for the expansivity of mercury may be found by a method independent of the expansion of the containing vessel. Give the necessary theory.

A silica bulb of negligible expansivity holds 340·0 g of mercury at 0 °C when full. Some steel balls are introduced and the remaining space is occupied at 0 °C by 255·0 g of mercury. On heating the bulb and its contents to 100 °C, 4·800 g of mercury overflow. Find the linear expansivity of the steel.

[Assume that the expansivity of mercury is 180 × 10^{-6} K^{-1}.] *(L.)*

14. Define the terms "apparent" and "absolute" expansivity of a liquid, and show how the former is found by means of a weight thermometer. A litre flask, which is correctly calibrated at 4 °C, is filled to the mark with water at 80 °C. What is the mass of water in the flask? [Linear expansivity of the glass of the flask = 8·5 × 10^{-6} K^{-1}; mean cubic expansivity of water = 5·0 × 10^{-4} K^{-1}.] *(O. and C.)*

5

Heat Capacities of Gases. Isothermal and Adiabatic Changes

Work done by gas

Consider a fixed mass of gas contained in a cylinder XLMY by a piston (fig. 5.1). If the gas expands from LM to CD, it does work against the external forces. We call this *external work*. For a small expansion LC, the pressure p of the gas may be considered constant. In this case the force exerted by the gas during expansion $= pA$, where A is the area of cross-section of the piston. If $LC = x$,

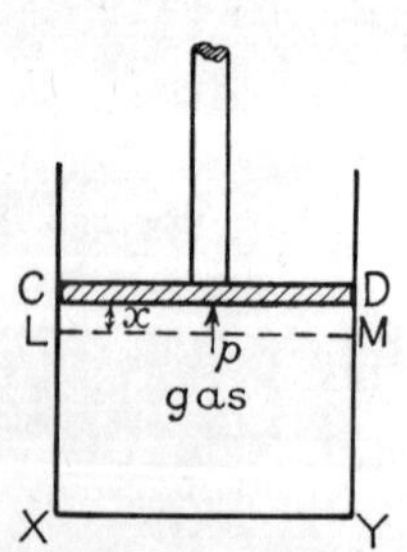

Fig. 5.1 Work done by gas on expansion

$$\therefore \text{ work done } W = \text{force} \times \text{distance} = pA \times x$$

$$= p \times \textit{volume change} \quad . \quad . \quad (1)$$

since $A \times x =$ volume LCDM. If the small volume change is represented by ΔV, then

$$W = p \, . \, \Delta V \quad . \quad . \quad . \quad . \quad (2)$$

Suppose a gas expands from a volume V_1 to a volume V_2 and that the pressure-volume changes occur along a graph PQ (fig. 5.2(i)). For a small change in volume, the work done $= p \, . \, \Delta V$, from (2), $=$ *area* between graph XY and volume-axis; this is shown shaded in fig. 5.2(i). It therefore follows that, when expanding from H (V_1) to K (V_2), the work done by the gas $=$ area between graph PQ and volume-axis $=$ area PQKH.

If the gas is compressed along a path QSP back to its original p, V values or original *state*, work is now done *on* the gas (fig. 5.2(ii)).

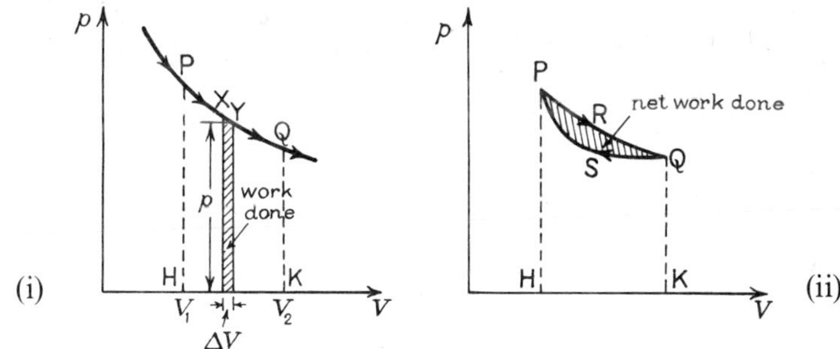

Fig. 5.2 Work done from p-V curves

The work done = area QSPHK. This is *less* than the work done *by* the gas along PRQ, and so, by subtraction

net work done on returning to P = area PRQSP.

This area is shown shaded in fig. 5.2 (ii). A process whereby a gas is returned to its original state is called a *cyclic process*, and it occurs continuously when a gas engine is operating (see p. 201).

Internal energy. Joule's experiments

If a gas expands without gaining any energy from an external source, the work done is derived from the store of *internal energy* of the gas. The internal energy U of a body is the sum of the kinetic energy of the thermal motion of the molecules and their potential energy due to intermolecular forces. The magnitude of U is not known, but this is not important as we are usually concerned with *changes* in U.

In 1845 Joule investigated the internal energy of a gas by allowing dry air under a pressure of 22 atmospheres to expand through a stop-cock S from a vessel A into an evacuated vessel B (fig. 5.3). No external work is then done by the gas, Joule argued, but if internal work is done, against the intermolecular attraction, the gas will be cooled as it then loses some energy. The vessels A, B were completely surrounded by water in another vessel, and although Joule used a very sensitive thermometer he was unable to detect a change in temperature of the water when the gas expanded into B. By means of this and other experiments, he concluded that

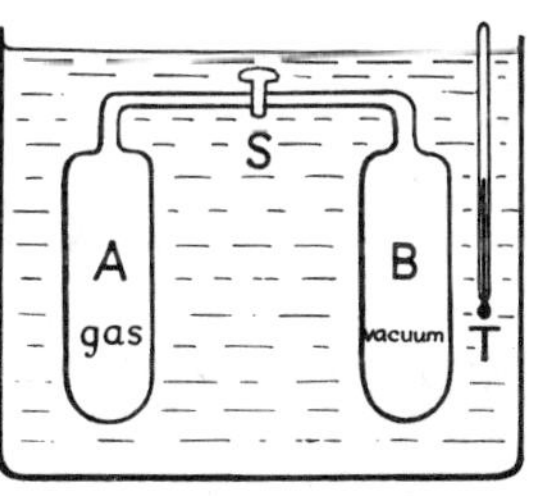

Fig. 5.3 Joule experiment on internal energy

the internal energy of a gas depends on its temperature and is independent of its volume. We know that this is true only for an *ideal gas*. Later experiments showed that, for real gases such as air, there is a drop of temperature on expansion (p. 145).

As we deal often with an ideal gas, it would be useful to list its main properties:

(1) It obeys Boyle's and Charles' laws, so that the cubic expansivity (volume coefficient) is equal to the pressure coefficient (p. 48).

(2) The relation $p = \frac{1}{3}\rho\overline{c^2}$, derived from the kinetic theory of gases, applies to an ideal gas, since intermolecular forces are neglected in deriving the relation (p. 60).

(3) The internal energy is independent of volume and depends only on the temperature of the gas.

First law of thermodynamics

Consider an ideal gas confined in a metal cylinder by a piston. If a quantity of heat is supplied to the gas, some of the energy is used in pushing back the piston or doing external work. The rest of the energy (heat) is used to raise the internal energy, which is the sum of the kinetic and potential energies of all the molecules. Hence, from the principle of the conservation of energy,

$$\delta Q = \delta U + \delta W$$

where δQ is the amount of heat supplied, δU is the increase in internal energy and δW is the external work done. The principle of the conservation of energy in heat and mechanical energy exchanges is also known as the *first law of thermodynamics*.

If no heat from an outside source is supplied to an ideal gas when it expands, the external work is equal to the loss in internal energy of the gas. The gas temperature thus falls. If the gas is insulated thermally and suddenly compressed, it gains internal energy equal to the external work done on it. The temperature of the gas thus rises.

Specific heat capacities of a gas

The specific heat capacity of a gas is the heat required to raise the temperature of unit mass by 1 °C. We have just seen that part of the heat supplied to the gas may be used to do external work as the gas expands. Now the pressure-volume changes of a gas may take place in many different ways. Consequently the specific heat capacity of a gas may have a large number of different values.

The specific heat capacity of a gas is therefore measured at constant volume or at constant pressure:

The *specific heat capacity of a gas at constant volume* c_V is the amount of heat required to raise the temperature of unit mass by 1 °C when the volume is constant throughout.

The *specific heat capacity of a gas at constant pressure* c_p is the amount of heat required to raise the temperature of unit mass by 1 °C when the pressure is constant throughout.

The SI unit of c_V or c_p is J kg^{-1} K^{-1}. Some values of specific heat capacities (also called "principal specific heats") are given in the table below.

	c_V (kJ kg^{-1} K^{-1})	c_p (kJ kg^{-1} K^{-1})	$\gamma (= c_p/c_V)$
hydrogen	10·0	14·1	1·41
oxygen	0·65	0·91	1·40
argon	0·31	0·52	1·67
air	0·72	1·00	1·39

Relation between specific heat capacities

Consider 1 kg of a gas contained in a cylinder by a piston (fig. 5.4 (i)). If the piston is kept in position at AB while the gas is heated, the volume XABY of the gas remains constant. The heat supplied to

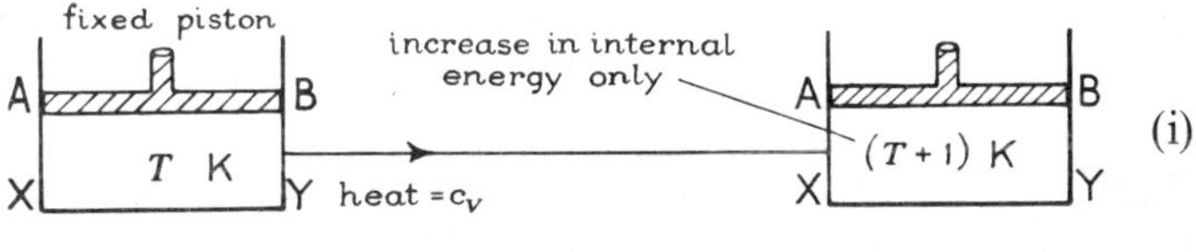

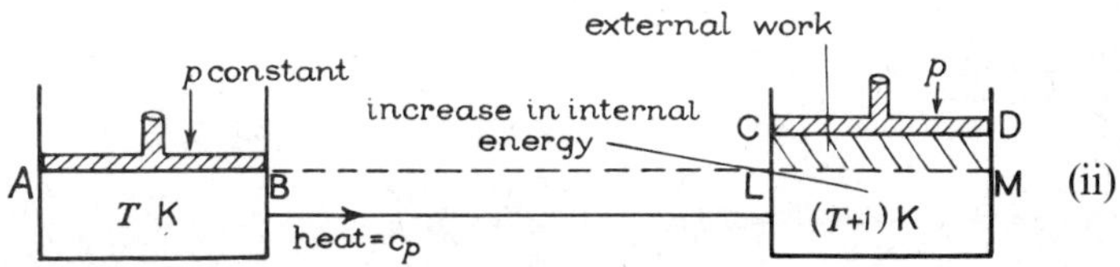

Fig. 5.4 Specific heat capacities of gas

the gas when its temperature rises by 1 °C is by definition c_V the specific heat capacity of the gas at constant volume. Note that in this case all the heat is used in increasing the kinetic energy of the molecules, that is, in increasing the internal energy of the gas. c_V is thus the increase in internal energy of unit mass of a gas per °C temperature rise. For an ideal gas c_V is independent of the actual volume of gas.

Suppose, however, that the gas is heated while the pressure p exerted by the piston is kept constant. The piston is then pushed back (fig. 5.4(ii)), *and some of the heat supplied is used in doing external work.* The remainder of the heat is used in raising the temperature of the gas, that is in increasing the internal energy of the gas.

Suppose that the piston is pushed back to CD when the temperature of the gas at constant pressure increases by 1 °C. The heat supplied to the gas is then by definition c_p, the specific heat capacity. From what has been stated before, it should be evident that c_p can be analysed into two parts:

(1) The heat required to raise the temperature of 1 kg of the gas by 1 °C—this is c_V, the specific heat capacity at constant volume, since the increase in internal energy per °C temperature rise is c_V.

(2) The heat required to do the external work W in pushing the piston back from its original position LM to its final position CD. It therefore follows that

$$c_p = c_V + W$$

The work done W can easily be calculated. On p. 98 it was shown that $W = p \times$ *volume change*. Suppose V_1 is the original volume of the gas and V_2 the final volume. Then since p is constant in this case,

$$W = p(V_2 - V_1) = pV_2 - pV_1$$

But $pV_1 = RT$, where R is the gas constant and T is the absolute temperature of the gas initially, and $pV_2 = R(T+1)$, since the temperature has risen 1 °C. Hence

$$W = pV_2 - pV_1 = R(T+1) - RT = R$$

$$\therefore c_p = c_V + R$$

or

$$c_p - c_V = R$$

The difference in the specific heat capacities of an ideal gas is thus equal to R, the gas constant per unit mass. If a molar mass of the gas is considered instead of unit mass, a similar relation is obtained. Thus

$$C_p - C_V = \mathbf{R}$$

where C_p, C_V represent molar heat capacities and $\mathbf{R}$ represents the molar gas constant (8·3 J mol^{-1} K^{-1}).

It can now be seen that c_p is always greater than c_V; the difference is the work done when the gas expands. In calculating the work, we were concerned only with the external work done (p. 102). Now experiment shows that the molecules of a real gas have some attraction for each other (p. 145). When a real gas expands, therefore, it does work against the intermolecular forces, called *internal work*, in addition to external work. The relation $c_p - c_V = R$ has been derived without taking into account the internal work and is an ideal gas relation. It is an approximate relation for real gases.

Example

At atmospheric pressure, the specific heat capacity at constant pressure of dry air is 1·00 kJ kg^{-1} K^{-1}. If the density of air at 0 °C and 10^5 N m^{-2} pressure is 1·29 kg m^{-3}, calculate a value for c_V.

$$\text{The gas constant per kg, } R = \frac{pV}{T} = \frac{10^5 \times 1/1{\cdot}29}{273}\ \text{J kg}^{-1}\,\text{K}^{-1}$$

$$= \frac{100 \times 1/1{\cdot}29}{273}\ \text{kJ kg}^{-1}\,\text{K}^{-1}$$

$$= 0{\cdot}28\ \text{kJ kg}^{-1}\,\text{K}^{-1}$$

$$\therefore c_V = c_p - R = 1{\cdot}00 - 0{\cdot}28 = 0{\cdot}72\ \text{kJ kg}^{-1}\,\text{K}^{-1}$$

Equipartition of energy. Energy of molecules

The molecules in a gas can move in one of three mutually perpendicular directions, that is, each molecule has three *degrees of freedom* of translational motion. A monatomic gas has only translational motion and hence three degrees of freedom. A diatomic or triatomic molecule has rotational and vibrational motion. They have therefore more degrees of freedom than a monatomic molecule. Each degree of freedom is associated with an amount of *energy*, and the *principle of equipartition of energy* states that *the same amount of energy is associated with each degree of freedom.*

On p. 59 it is shown on the simple kinetic theory of gases that the translational kinetic energy U per mole of a monatomic gas is given by $U = 3\mathbf{R}T/2$, where $\mathbf{R}$ is the molar gas constant. A monatomic gas has three degrees of freedom. Hence, from the principle of equipartition of energy,

$$\text{average energy per degree of freedom} = \tfrac{1}{3} \times \tfrac{3}{2}\mathbf{R}T = \tfrac{1}{2}\mathbf{R}T$$

If N_A is the number of molecules per mole, each molecule will have an average amount of energy per degree of freedom given by

$$\tfrac{1}{2}\frac{\mathbf{R}}{N_A}T = \tfrac{1}{2}kT$$

where k = *Boltzmann's constant* $= \mathbf{R}/N_A$
$= 8{\cdot}3\ \text{J mol}^{-1}\ \text{K}^{-1}/6{\cdot}02 \times 10^{23}\ \text{mol}^{-1} = 1{\cdot}38 \times 10^{-23}\ \text{J K}^{-1}$.

A simple mechanical model of a diatomic gas is a "dumb-bell", with each atom at one end of the bar (fig. 5.5 (i)). Diatomic molecules have rotational degrees of freedom in addition to translational degrees of freedom. In general, the rotational motion can be resolved about two axes perpendicular to the axis of the molecule. Hence there are

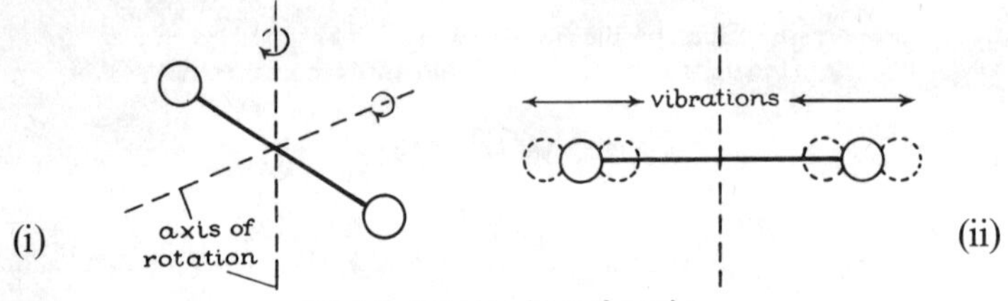

Fig. 5.5 Rotational and vibrational degrees of freedom

two degrees of rotational freedom in addition to three degrees of translational freedom (fig. 5.5 (i)). With a total of five degrees of freedom, and assuming the average kinetic energy in each degree of freedom is equal to $\tfrac{1}{2}kT$, the total kinetic energy of translation and rotation is $\tfrac{5}{2}kT$. The total kinetic energy per mole, with N_A molecules, is $\tfrac{5}{2}N_AkT$ or $\tfrac{5}{2}\mathbf{R}T$. If the molecules have vibrational energy in addition, this also contributes two degrees of freedom since the energy in vibration is partly potential and partly kinetic (fig. 5.5 (ii)). Thus the total kinetic energy per mole of the diatomic molecules is greater than $\tfrac{5}{2}\mathbf{R}T$.

Generally, with f degrees of freedom, the average kinetic energy per molecule is $\tfrac{1}{2}fkT$ and the total kinetic energy per mole is $\tfrac{1}{2}f\,\mathbf{R}T$.

Molar heat capacity

The *molar heat capacity* of a substance is the heat (energy) required to raise the temperature of one mole by one °C (1 K).

In the case of a *solid*, a molecule has three vibrational degrees of freedom, each having potential and kinetic energy (p. 103). Thus the average total energy per molecule $= 3 \times 2 \times \frac{1}{2}kT = 3kT$, and the average energy per mole $= 3\mathbf{R}T$. Hence the molar heat capacity = heat (energy) per K = 3**R**. Thus, on classical theory, the molar heat capacity of a solid is constant at all temperatures and has a value of $3 \times 8{\cdot}3$ or about 25 J mol^{-1} K^{-1} (*Dulong and Petit law*). In practice, this is roughly the case at ordinary and high temperatures only.

For a *monatomic gas*, the molar heat capacity at constant volume = the increase in internal energy per K $= \frac{3}{2}\mathbf{R} = 12{\cdot}5$ J mol^{-1} K^{-1}. This is in fair agreement in practice. For a *diatomic gas*, the molar heat capacity at constant volume $= \frac{7}{2}\mathbf{R}$, assuming it has 2 degrees of freedom of vibrational energy, 2 degrees of rotational energy, and 3 degrees of translational energy. The variation of heat capacity with temperature, obtained experimentally, is shown in fig. 5.6. At high temperatures the molar heat capacity is $\frac{7}{2}\mathbf{R}$ but classical theory fails at lower temperatures. Here the thermal energy $\frac{1}{2}kT$ is insufficient to excite the molecule into its rotational or vibrational energy levels, as derived from quantum theory.

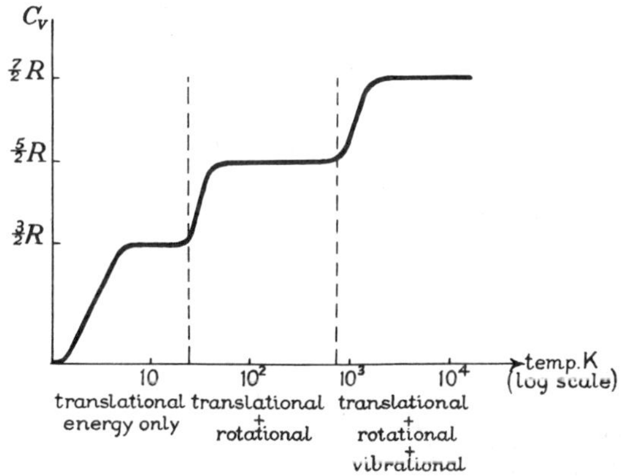

Fig. 5.6 Variation of molar heat capacity with temperature

The molar heat capacity tends to a very low value as the temperature approaches the absolute zero, and this is also the case for solids.

On classical theory, the free electrons in the atoms of metals should contribute energy additional to that of the vibrating atoms. Thus, since each electron has three degrees of freedom of translational energy, the total energy per mole $= 3\mathbf{R} + 3\mathbf{R}/2 = 9\mathbf{R}/2$. This does not agree with experimental results. In 1926 Fermi and independently

Dirac, two eminent mathematical physicists, applied a new type of statistics to the electrons in metals, based on the quantum theory and the exclusion principle of Pauli. This is outside the scope of the book, and we can only quote their results. They showed that the electrons are not a "gas" to which the classical (Maxwell-Boltzmann) statistics apply, that is, each electron does not contribute an amount of energy $\frac{1}{2}kT$ per degree of freedom. Fermi-Dirac statistics reveal, in fact, that electrons contribute a negligible amount to the specific heat at ordinary temperatures.

Debye has shown that the atomic vibration contribution to specific heats is proportional to T^3 at low temperatures for most crystals, including metals. This is the major contribution to the specific heat down to about 5 K. Below this temperature the electron contribution becomes comparable with that contributed by the atomic vibration.

Monatomic and diatomic gas molecules

Experimental results for the heat capacities of gases show what proportion of the total energy supplied to a gas at constant volume is transformed to translational kinetic energy of their molecules.

Argon, a monatomic gas of molar mass 40, has a specific heat capacity at constant volume c_V of about 0·315 J g^{-1} K^{-1} and hence a molar heat capacity C_V of 40 × 0·315 or 12·6 J mol^{-1} K^{-1}.

The molar gas constant **R** is about 8·3 J mol^{-1} K^{-1}. Hence the translational energy per mole of the molecules $= \frac{3}{2}\mathbf{R}T = \frac{3}{2} \times 8{\cdot}3\ T$ or 12·45 T J. Thus if the temperature of the gas is raised 1 °C, the translational energy increases by 12·45 J. But the total energy supplied $= C_V =$ 12·6 J. Hence it can be seen that, for the monatomic gas, practically the whole of the energy supplied is transformed to translational kinetic energy.

Oxygen, a diatomic gas of molar mass 32, has a specific heat capacity at constant volume c_V of about 0·65 J g^{-1} K^{-1} and hence a molar specific heat C_V of 32 × 0·65 or 20·8 J mol^{-1} K^{-1}. Now, from above, the translational energy per mole of the molecules increases by about 12·5 J per °C temperature rise. This is a fraction 12·5/20·8 or about $\frac{3}{5}$ of the total energy supplied. The rest of the energy is transformed to rotational and vibrational energy (p. 103).

Ratio of specific heat capacities

The specific heat capacity of a monatomic gas at constant volume $c_V =$ heat to raise the internal energy of unit mass of the gas by 1 °C or 1 K $= \frac{3}{2}R$, where R is the gas constant per unit mass (p. 105). Now $c_p - c_V = \mathrm{R}$ (p. 102).

$$\therefore\ c_p = R + c_V = \tfrac{5}{2}R.$$

$$\therefore \text{ ratio of specific heats, } \gamma = \frac{c_p}{c_V} = \frac{\frac{5}{2}R}{\frac{3}{2}R} = \frac{5}{3} = 1{\cdot}67$$

Experiments to measure γ (see p. 115) show that it has a value close to 1·67 for monatomic gases. This is experimental evidence in favour of the kinetic theory of gases.

With a diatomic gas, $c_V = \frac{5}{2}R$ if the vibrational energy is negligible compared with the translational and rotational energy. Assuming this is the case, then $c_p = \frac{7}{2}R$ and $\gamma = 7/5 = 1{\cdot}4$. Many diatomic gases have values of γ close to 1·4. If the vibrational energy is not negligible, $c_V = \frac{7}{2}R$ and $\gamma = 9/7 = 1{\cdot}3$ (approx). It can be seen that, with f degrees of freedom, $c_V = \frac{1}{2}fR$ and $c_p = (1 + \frac{1}{2}f)R$, so that $\gamma = 1 + 2/f$. It should be noted that "models" of molecules, such as that given previously for a diatomic molecule, provide only a guide to the behaviour of molecules when heated and must not be taken as literally true.

ISOTHERMAL AND ADIABATIC CHANGES

Isothermal expansion

In a steam engine, the vapour expands at constant temperature in a cylinder during one stage of the action; the same type of gas expansion occurs at stages in other gas engines. Changes of pressure and volume which take place at *constant temperature* are known as *isothermal* changes.

We can understand how an isothermal change may take place if we suppose that a quantity of gas is enclosed in a metal cylinder by a piston. If the piston is depressed *very slowly*, mechanical work is done on the gas and a little heat is produced. But the metal cylinder is a good conductor of heat, and the heat has time to escape from the gas through the metal. The gas therefore remains at the same temperature as the cylinder, which may be room temperature, for example, while the gas contracts. This is an example of an isothermal change.

Similarly, suppose that the piston in the cylinder is released very slowly, so that the gas expands. This time the gas does work, and the energy is taken from the gas itself, which therefore decreases slightly in temperature. Heat now flows from the surroundings through the metal cylinder to the gas, however, and the temperature of the gas remains unaltered; an isothermal expansion has thus taken place.

Since an isothermal change is one which takes place at constant temperature, the pressure-volume changes obey Boyle's law. Thus $pV = k$, where k is a constant equal to RT. T is the absolute value of the constant temperature, and R is the gas constant per unit mass,

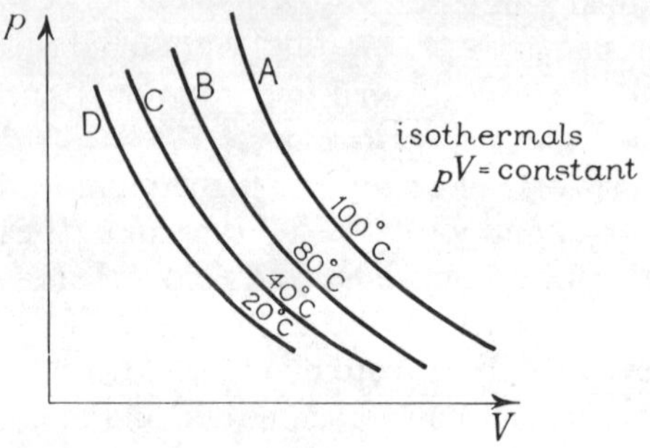

Fig. 5.7 Isothermals of gas

assuming this is the mass of gas. If the temperature of the gas is kept constant at 100 °C, the curve A is obtained when the values of pressure (p) and volume (V) are plotted (fig. 5.7). Similarly shaped curves, B, C, D, are obtained at other constant temperatures, 80 °C, 40 °C, 20 °C respectively. These curves are all known as *isothermals* of the gas under investigation.

Reversible and irreversible changes

An important concept in the subject of thermodynamics (p. 202) is that of a *reversible change*. Consider a gas contained in a cylinder by a piston which is very light and frictionless, and suppose the gas has a pressure p. Then, with the piston in equilibrium, the external pressure on the piston is also equal to p. Now suppose the external pressure is increased infinitely slowly by a small amount δp. In this ideal case, as the pressure of the gas rises slowly, the pressure at every stage is equal and opposite to the external pressure. At the end of the process, the pressure of the gas has increased to $p + \delta p$ say, the volume has decreased to $V - \delta V$, and the temperature may have changed to $T + \delta T$. The state of the gas throughout this ideal change is described only by the three quantities p, V, T, for example, by $pV = RT$ or by $(p + a/V^2)(V - b) = RT$. If the piston is now raised very slightly and infinitely slowly, the gas will return to its original values of pressure, volume and temperature, retracing exactly all the values it had while the piston was lowered. Throughout the change the three variables are again related only by the equation of state of the gas.

This ideal change is called a *reversible change*, or a *quasistatic change in which no dissipation of energy occurs*, for example, to overcome friction. It may be defined as one in which the pressure of the gas differs only by an infinitesimal amount from the external pressure at every stage, and in which any change in volume of the gas takes place very slowly, without dissipative effects due to friction or viscosity. In this case only the smallest change of conditions in the opposite direction is needed to make the process retrace exactly its previous values of p and V. A reversible change is impossible to make in practice, but it represents an ideal change with which practical changes can be compared.

If the piston containing the gas is heavy, there will always be a finite difference between the pressure inside and outside the gas. The net pressure when the piston is moved down may then produce acceleration and hence turbulence in the gas, which may set up a non-uniform temperature distribution. This is an example of an *irreversible process*; a reverse change in direction of the piston would produce turbulence again. Natural processes appear to be irreversible.

Reversible isothermal changes. Work done

Suppose a cylinder, made of perfectly conducting material, contains an ideal gas confined by a light piston, and is placed in contact with a large reservoir of heat at a constant absolute temperature T. If the gas expands very slowly, so that a reversible change occurs isothermally, the heat δQ entering the gas from the reservoir is equal to the external work done by the gas, as no work is done against friction and the piston is light. Here $\delta Q = p\,.\,\delta V$, where p has the value given by $pV = RT$ throughout the change. Conversely, in a reversible isothermal contraction of an ideal gas, the work done on the gas $= p\,.\,\delta V = \delta Q$, where δQ is the heat given up to the reservoir at the constant temperature concerned.

If the isothermal expansion of a gas from a volume V_1 to a volume V_2 is reversible, then, from above,

$$\text{work done } W = \int_{V_1}^{V_2} p\,.\,dV$$

where $p = RT/V$.

$$\therefore\ W = RT\int_{V_1}^{V_2} \frac{dV}{V} = RT\log_e\left(\frac{V_2}{V_1}\right)$$

Adiabatic expansion

In addition to isothermal changes, the gases in an engine expand and contract under such conditions that *no heat enters or leaves them.* The corresponding pressure-volume changes are then said to take place *adiabatically*. An adiabatic change is defined as one during which no heat enters or leaves the system concerned; in other words, $\delta Q = 0$ in the first law of thermodynamics (p. 100).

We can understand how an adiabatic change can take place if we again consider a gas contained inside a cylinder fitted with a piston; but this time, we shall suppose that the outside of the cylinder has been surrounded by insulating material, such as cotton-wool, and that the piston is also made of insulating material. If the piston is depressed, work is done on the gas, and an equivalent amount of heat is produced. Unlike the case of an isothermal change, however, no heat escapes from the gas because the container and piston are insulated; and the temperature of the gas rises. This is an example of an adiabatic contraction. If the piston is raised, work is done by the gas, and the equivalent heat is taken from the gas itself, which is therefore cooled. No heat, however, enters or leaves the gas while it expands, and this is an example of an adiabatic expansion. A sudden compression or expansion of a gas is *initially* adiabatic because there is then no time for heat to enter or leave the gas.

Reversible adiabatic pressure-volume changes

When a gas undergoes a reversible adiabatic change the pressure-volume changes obey the law

$$pV^{\gamma} = \text{constant}$$

where γ is the ratio of the specific heat of the gas at constant pressure to the specific heat at constant volume, i.e. $\gamma = c_p/c_V$. This relation between p and V is proved later (see p. 114). For monatomic gases, such as argon and helium, at normal temperatures, $\gamma = 1{\cdot}67$ (approximately; for most diatomic gases and air, $\gamma = 1{\cdot}4$ approx (see p. 107).

We now illustrate how new pressures and volumes of gases are calculated when reversible adiabatic changes occur.

Example

Suppose the pressure of 100 cm³ of air is 75 cm of mercury, and that the gas is compressed adiabatically and reversibly so that its volume becomes 25 cm³. Then, since $pV^{\gamma} =$ constant and $\gamma = 1{\cdot}4$ (approx) for air, we have

$$p \times 25^{1\cdot 4} = 75 \times 100^{1\cdot 4}$$

where p is the new pressure of the gas.

$$\therefore p = \frac{75 \times 100^{1\cdot 4}}{25^{1\cdot 4}}$$

Now $$\log p = \log 75 + 1{\cdot}4 \log 100 - 1{\cdot}4 \log 25 = 2{\cdot}7180$$

$$\therefore p = 522 \text{ cm mercury}$$

The calculation is slightly more difficult if the new volume of a gas is required. Suppose, for example, that the volume of a given mass of air is 80 cm³ at a pressure of 60 cm of mercury, and that the gas expands adiabatically to a final pressure of 45 cm mercury. The new volume V of the air is then given by

$$45 \times V^{1\cdot 4} = 60 \times 80^{1\cdot 4}$$

since $pV^{1\cdot 4}$ is a constant.

$$\therefore V^{1\cdot 4} = \frac{60 \times 80^{1\cdot 4}}{45}$$

Taking logs to the base 10 on both sides, we have

$$1{\cdot}4 \log V = \log 60 + 1{\cdot}4 \log 80 - \log 45$$

$$\therefore \log V = (\log 60 + 1{\cdot}4 \log 80 - \log 45)/1{\cdot}4 = 1{\cdot}9924$$

$$\therefore V = 98 \text{ cm}^3$$

Adiabatic curves. Temperature variation in reversible adiabatic changes

Suppose the pressure p_1 and volume V_1 of a gas are represented by the point Q in fig. 5.8. If the absolute temperature of the gas is T_1, the point Q lies on the isothermal curve corresponding to this

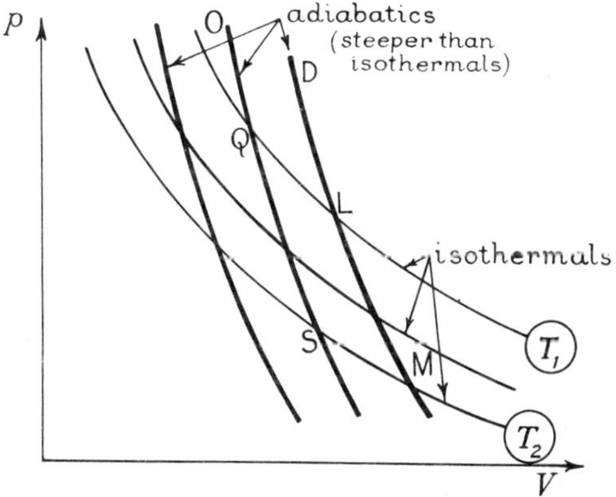

Fig. 5.8 Isothermals and adiabatics

temperature. Now suppose that the gas undergoes an adiabatic expansion; the temperature of the gas then decreases to an absolute value, T_2 say, and if pressure and volume reach a point corresponding to S, the latter lies on the isothermal of T_2 (fig. 5.8). OQS is thus the p–V curve obeying the relation $pV^{\gamma} =$ constant. By similar reasoning DLM is the p–V curve if the pressure and volume of the same mass of gas are originally represented by the point L and an adiabatic change takes place. Thus *adiabatics*, as the pV^{γ} curves are called, are generally steeper than isothermals of the same mass of gas, as fig. 5.8 illustrates.

We are now in a position to find how the temperature varies in an adiabatic change. Suppose that 1 kg of a gas has a pressure p_1, a volume V_1, and an absolute temperature T_1, corresponding to Q (fig. 5.8); and that after an adiabatic expansion, during which the pressure-volume changes take place along OQS, the gas has a new pressure p_2, a new volume V_2, and a new absolute temperature T_2, corresponding to S. Then, since Q is on the isothermal for T_1, $p_1V_1 = RT_1$; also, since S is on the isothermal for T_2, $p_2V_2 = RT_2$.

$$\therefore \frac{p_1V_1}{T_1} = \frac{p_2V_2}{T_2} \qquad \text{(i)}$$

But Q and S are both on the adiabatic curve OQS, which has the equation $pV^{\gamma} =$ constant.

$$\therefore p_1V_1^{\gamma} = p_2V_2^{\gamma} \qquad \text{(ii)}$$

To eliminate p_1 and p_2, we divide (ii) by (i). Hence

$$T_1V_1^{\gamma-1} = T_2V_2^{\gamma-1}$$

Thus, in general, for an adiabatic change

$$TV^{\gamma-1} = \text{constant}$$

Eliminating V from $pV = RT$ and $pV^{\gamma} =$ constant, we obtain

$$\frac{p^{\gamma-1}}{T^{\gamma}} = \text{constant}$$

Work done in reversible adiabatic expansion

If the gas volume increases from V_1 to V_2, then, if $pV^{\gamma} = c$, a constant,

$$\text{work done,} \quad W = \int_{V_1}^{V_2} p\,dV = \int_{V_1}^{V_2} \frac{c}{V^{\gamma}}\,dV$$

$$= \frac{1}{\gamma-1}\left[\frac{c}{V_1^{\gamma-1}} - \frac{c}{V_2^{\gamma-1}}\right] = \frac{1}{\gamma-1}\left[\frac{p_1V_1^{\gamma}}{V_1^{\gamma-1}} - \frac{p_2V_2^{\gamma}}{V_2^{\gamma-1}}\right]$$

$$= \frac{1}{\gamma-1}(p_1V_1 - p_2V_2) = \frac{R}{\gamma-1}(T_1 - T_2)$$

The work done is proportional to $(T_1 - T_2)$, the temperature fall. This can also be seen from the fact that *the work done in expansion is equal to the decrease in internal energy* of the gas, since no heat is supplied to the system. The decrease in internal energy $-\delta U_1 = -mc_V . \delta T$ generally or $mc_V(T_1 - T_2)$, where c_V is the specific heat capacity at constant volume. Hence the work done $\propto (T_1 - T_2)$, as before.

Examples

1. Suppose that a mass of air at a temperature of 40 °C and pressure 70 cm of mercury is compressed adiabatically and reversibly until its volume is halved. If its original volume is V, its new absolute temperature is given by

$$T \times (\tfrac{1}{2}V)^{\gamma-1} = 313 \times V^{\gamma-1}$$

since 40 °C = 313 K.

$$\therefore T = 313 \times \left(\frac{V}{\frac{1}{2}V}\right)^{\gamma-1} = 313 \times 2^{\gamma-1}$$

Assuming $\gamma = 1{\cdot}4$ for air, $T = 313 \times 2^{0\cdot4} = 413$ K. Hence the new temperature of the gas = 413 − 273 = 140 °C.

The new pressure p of the gas is easily obtained, as pV^γ = constant.

Thus $$p(\tfrac{1}{2}V)^{1\cdot4} = 70(V)^{1\cdot4}, \quad \text{i.e. } p = 70 \times \left(\frac{V}{\frac{1}{2}V}\right)^{1\cdot4} = 70 \times 2^{1\cdot4}$$

$$= 185 \text{ cm of mercury.}$$

2. How are the pressure and volume of a fixed mass of gas related during reversible isothermal expansion and during reversible adiabatic expansion?

22·4 litres of air at 15 °C and 76 cm of mercury pressure, weighing 27·3 g, expand adiabatically until the volume has increased by 50 per cent. What will be the final pressure and temperature?

By considering the internal energy of the gas find how much work is done against external pressure during the expansion [$c_V = 0{\cdot}71$; $c_p = 1{\cdot}00$ kJ kg^{-1} K^{-1}]. (*L.*)

First part.—In an isothermal expansion, pV = constant; in an adiabatic expansion, pV^γ = constant.

Second part.—As 22·4 litres is the original volume, the final volume is 33·6 litres. Since $\gamma = c_p/c_V$, $\gamma = \dfrac{1{\cdot}00}{0{\cdot}71} = 1{\cdot}4$ (approx). Applying pV^γ = constant, we have

$$p \times 33{\cdot}6^\gamma = 76 \times 22{\cdot}4^\gamma$$

where p is the final pressure.

$$\therefore p = 76\left(\frac{22{\cdot}4}{33{\cdot}6}\right)^{1\cdot4} = 76 \times \left(\frac{2}{3}\right)^{1\cdot4}$$

$$\therefore \log p = \log 76 + 1{\cdot}4 \log 2 - 1{\cdot}4 \log 3 = 1{\cdot}6343$$

$$\therefore p = 43{\cdot}1 \text{ cm of mercury}$$

Since $TV^{\gamma-1}$ = constant for an adiabatic change (p. 112), and 15 °C = 288 K, we have

$$T(33{\cdot}6)^{1\cdot4-1} = 288(22{\cdot}4)^{1\cdot4-1}$$

where T is the final temperature.

$$\therefore T = 288 \times \left(\frac{22 \cdot 4}{33 \cdot 6}\right)^{0 \cdot 4}$$

$$= 288 \times (\tfrac{2}{3})^{0 \cdot 4}$$

$$\therefore \log T = \log 288 + 0 \cdot 4 \log 2 - 0 \cdot 4 \log 3 = 2 \cdot 3890$$

$$\therefore T = 245 \text{ K}$$

$$\therefore \text{temperature in } ^\circ\text{C} = 245 - 273 = -28$$

Third part.—Since no external heat is supplied to a gas in an adiabatic expansion, the change in the internal energy of the gas = the external work done (p. 102). But the change in the internal energy = mass of gas × c_V × temperature change.

$$\therefore \text{internal energy change} = 27 \cdot 3 \times 0 \cdot 71 \times [15 - (-28)]$$

$$= 833 \text{ J}$$

Proof of pV^γ constant

Suppose that an ideal gas expands adiabatically and reversibly (p. 112). Then no heat enters or leaves the gas. Now from the first law of thermodynamics, $\delta Q = \delta U + \delta W = \delta U + p . \delta V$ (p. 100). Since $\delta Q = 0$, then $\delta U + p . \delta V = 0$. But $\delta U = c_V . \delta T$, where δT is the small change in temperature of unit mass of the gas (p. 113).

$$\therefore c_V . \delta T + p . \delta V = 0$$

$$\therefore \delta T = -\frac{p . \delta V}{c_V} \quad . \quad . \quad . \quad . \quad . \quad . \quad \text{(i)}$$

Now if we deal with 1 kg of gas, $pV = RT$ always applies to the pressure, volume, and absolute temperature values of the gas. Taking small changes in both sides of the gas equation, and noting that pV is a product of two variables, we have

$$p . \delta V + V . \delta p = R . \delta T$$

But $c_p - c_V = R$, assuming R is expressed in the same units as c_p and c_V.

$$\therefore p . \delta V + V . \delta p = (c_p - c_V)\delta T \quad . \quad . \quad . \quad \text{(ii)}$$

Substituting for δT from (i) in (ii),

$$\therefore p . \delta V + V . \delta p = -\frac{(c_p - c_V) . p\delta V}{c_V} = -\frac{c_p}{c_V} p . \delta V + p . \delta V$$

$$\therefore V . \delta p = -\frac{c_p}{c_V} p . \delta V = -\gamma p . \delta V$$

where $\gamma = \frac{c_p}{c_V}$.

$$\therefore \frac{\delta p}{p} = -\gamma \frac{\delta V}{V}$$

This is the relation which holds between small changes of p and V in a reversible adiabatic change. To find the relation between the actual values of p and V, this relation must be integrated.

$$\therefore \int \frac{dp}{p} = -\gamma \int \frac{dV}{V}$$

$$\therefore \log_e p = -\gamma \log_e V + a$$

where a is a constant depending on the initial conditions of pressure and volume.

$$\therefore \log_e p + \gamma \log_e V = a, \quad \text{i.e. } \log_e p + \log_e V^{\gamma} = a$$

$$\therefore \log_e (pV^{\gamma}) = a$$

$$\therefore pV^{\gamma} = e^a$$

But e^a is a constant,

$$\therefore pV^{\gamma} = \text{constant}$$

Determination of γ. Clément and Désormes' experiment

Clément and Désormes designed a simple method of measuring γ, the ratio of the specific heat capacities of a gas. A lagged large vessel M was fitted with a tap at A and a manometer N, containing oil. By means of a pump connected to P, the pressure of the gas, say air, in M was increased until there was an appreciable difference in level between the oil columns in N (fig. 5.9 (i)). The pressure p_1 of the air

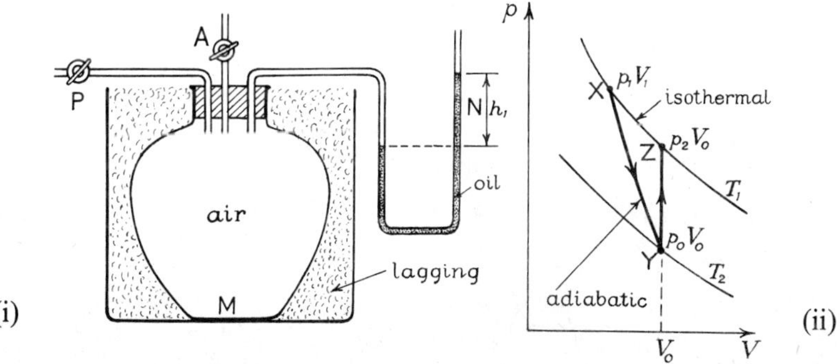

Fig. 5.9 Clément and Désormes' experiment and theory

in M is then slightly above atmospheric pressure p_0. The tap A is now quickly opened and closed, when some of the air escapes from M, and after conditions are steady, the final new pressure p_2 is read from the manometer N.

Consider the mass of air *left* in M, and suppose that it had a volume V_1 before the tap A was opened. When A is opened, the volume of this mass of air increases rapidly to V_0, the volume of the vessel M; and since the expansion takes place rapidly and M is lagged, the change in pressure and volume occurs under *adiabatic* conditions. The air is then cooled from room temperature T_1 to a temperature T_2, as shown by XY in fig. 5.9(ii). Thus, from XY,

$$p_1 V_1^{\gamma} = p_0 V_0^{\gamma} \quad . \quad . \quad . \quad . \quad . \quad . \quad \text{(i)}$$

After a time, when the air left in M has completely settled down, the temperature of this air becomes the same as that of the surroundings. We shall suppose its pressure is then p_2, so that the air returns along YZ to the temperature T_1 as shown in fig. 5.9 (ii). Since Boyle's law is applicable to this fixed mass of air, which has finally a volume V_0 at T_1,

$$p_1 V_1 = p_2 V_0 \quad . \quad . \quad . \quad . \quad . \quad . \quad . \quad . \quad . \quad . \quad \text{(ii)}$$

From (ii), $\dfrac{V_0}{V_1} = \dfrac{p_1}{p_2}$. But, from (i), $\dfrac{V_0^{\gamma}}{V_1^{\gamma}} = \dfrac{p_1}{p_0}$

$$\therefore \left(\frac{p_1}{p_2}\right)^{\gamma} = \frac{p_1}{p_0} \quad . \quad . \quad . \quad . \quad . \quad . \quad . \quad . \quad . \quad \text{(iii)}$$

In practice, $p_1 = p_0 + h_1$, where h_1 is the small initial difference in levels of oil in the manometer, and $p_2 = p_0 + h_2$, where h_2 is the small final difference in levels. From (iii),

$$\gamma \log \left(\frac{p_1}{p_2}\right) = \log \left(\frac{p_1}{p_0}\right)$$

$$\therefore \gamma = \frac{\log (p_1/p_0)}{\log (p_1/p_2)} = \frac{\log (1 + h_1/p_0)}{\log [1 + (h_1 - h_2) p_0]}$$

to a good approximation. Since $\log (1 + x) = x$, when x is small, it follows that

$$\gamma = \frac{h_1/p_0}{(h_1 - h_2)/p_0} = \frac{h_1}{h_1 - h_2}$$

In this way γ may be quickly found.

In Clément and Désormes experiment, the gas undergoes a sudden adiabatic expansion and oscillations occur in the neck of the flask.

Owing to the frictional forces on the gas, heat is produced and so the change is irreversible. The equation $pV^{\gamma} =$ constant is therefore obeyed only approximately. More accurate methods for γ involve the measurement of the temperature change after adiabatic expansion of a gas.

Measurement of c_V. Joly's differential steam calorimeter

Joly designed a calorimeter for measuring the specific heat capacity of a gas at constant volume, known as a *differential steam calorimeter*. Figure 5.10 illustrates the principle of the experiment. Two identical copper spheres A and B are suspended in a chamber from the scale-pans of a balance, and while A is filled with the gas whose specific

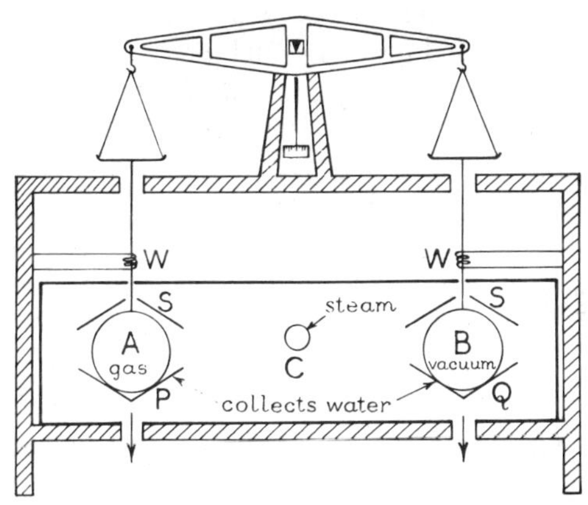

Fig. 5.10 Measurement of c_V—Joly differential steam calorimeter

heat is required, B is completely evacuated. P and Q are two pans beneath the respective spheres. Initially a balance is obtained by adding weights to the scale-pan to which B is attached, and when suitable corrections are made, the mass of the gas in A is known.

Steam is now passed into the chamber through a pipe at C, and after a time, when conditions are steady, *more steam is observed to have condensed on* A *than on* B, the scale-pans P and Q collecting the water formed. The balance is then restored, and the extra mass of steam M condensed on A is now known.

Theory.—The temperature, t °C say, in the chamber is that of the steam. The two spheres A and B, and the gas in A, have thus been raised to a temparature of t °C from an initial temperature t_1 °C say. Now since A and B are identical spheres, and A contains a gas while B contains a vacuum, more heat is required to raise the temperature of A and its contents to t °C, than to raise the temperature of B to t °C. The difference is the heat required to raise the temperature of a mass m of the gas at constant volume from t_1 °C to t °C. But this is the heat given out by a mass M of steam at t °C when it condenses to water at the same temperature, which is Ml joules, where l is the specific latent heat of steam. Hence

$$mc_V(t - t_1) = Ml$$

$$\therefore c_V = \frac{Ml}{m(t - t_1)}$$

Thus c_V can be calculated from a knowledge of M, m, t, t_1, and l.

In fig. 5.10 SS represent shields which prevent the water condensing on the roof of the chamber from being collected by P and Q. WW represents electrical heaters, which prevent drops forming round the wire attached to the scale-pans; such drops would hinder the free movement of the scale-pans. Corrections are necessary for the small expansion of the copper sphere containing the gas, as its temperature rises, thus altering the volume of the gas slightly; and also for a slight difference in the heat capacity of the two spheres. It will be noted that the use of the two identical spheres eliminates the necessity of considering the heat capacity of the sphere containing the gas; it is also the reason for the name "differential" steam calorimeter, as the *difference* in the masses of the steam condensed on the two spheres is used.

Measurement of c_p by continuous flow

An accurate method of determining the specific heat capacity of a gas at constant pressure c_p is that of the continuous-flow (electrical) method, used in measuring the specific heat capacity of a liquid (see p. 16).

The apparatus used is illustrated in fig. 5.11. A steady flow of gas at constant pressure is introduced into the glass tube at A, and its temperature is measured by the platinum resistance thermometer T_1. The gas then flows through a coiled tube, represented by C, which steadies the flow, and issues slowly past an electrical heating coil D towards a gauze G, whose purpose is to mix the gas so that the whole of it is at a uniform temperature. This temperature is measured by the platinum resistance thermometer T_2, and the gas issues at B.

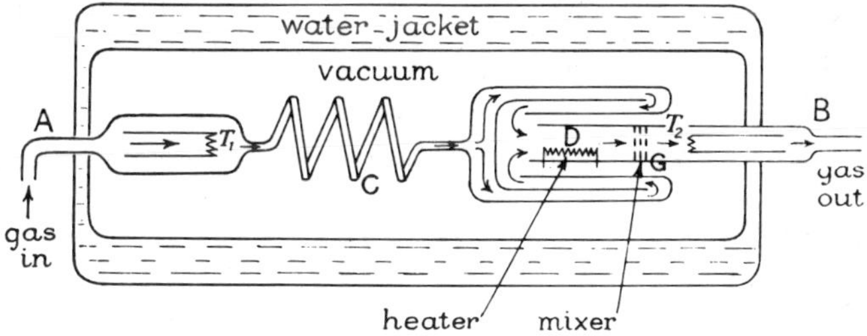

Fig. 5.11 Electrical method for c_p

Suppose m is the mass of the gas in kg flowing through the tube per second, t is its rise in temperature, and I, V are the current in amperes in the heating coil and the potential difference across it in volts. The heat in joules supplied per second by the current is then IV and, neglecting any heat lost, is equal to the heat per second gained by the gas.

$$\therefore mc_p t = IV$$

$$\therefore c_p = \frac{IV}{mt}$$

A pressure gauge connected to the vessel containing the gas is used to measure the mass $\overline{m}$ of gas flowing through the tube. From $m = pV/RT$ (p. 51), $\overline{m} = (p_1 - p_2)V/RT$, where p_1, p_2 are the initial and final pressures, V is the volume of the vessel, T is the absolute temperature and R is the gas constant per unit mass. A vacuum jacket serves to diminish loss of heat by cooling (fig. 5.11); the very small amount of heat lost can be found by repeating the experiment, as explained on p. 17. The continuous-flow electrical method has advantages listed on p. 18.

EXERCISE 5

Fill in the missing words or symbols in questions 1–4:

1. For an ideal gas, the internal energy is .... of the volume.

2. $c_p - c_V =$ .... for an ideal gas.

3. For a real gas, work is done against .... when expansion occurs.

4. For unit mass of a monatomic ideal gas, ... $= \frac{3}{2}R$;
the energy per degree of freedom = ...

5. If $c_p = 1{\cdot}0$ kJ kg^{-1} K^{-1} for air and 1 kg of air at s.t.p. (0 °C and $1{\cdot}013 \times 10^5$ N m^{-2} pressure) has a volume of $0{\cdot}775$ m^3, calculate a value for c_V.

6. What is meant by *reversible isothermal expansion* and by *reversible adiabatic expansion*? What is the relation between pressure and volume in each case. A gas of volume 800 cm^3 and 720 mmHg pressure expands isothermally to a volume of 900 cm^3. Find the new pressure.

7. A gas expands from a volume of 100 cm^3 to 250 cm^3 under a constant pressure of 2 atmospheres. What external work has been done by the gas if 1 atmosphere $=1{\cdot}013\times10^5$ newton metre^{-2}. Has the gas gained or lost (i) heat during the expansion, (ii) internal energy, assuming it is an ideal gas?

8. The volume of a gas at 15 °C and 80 cm mercury pressure is 100 cm^3. The gas is compressed to a volume of 60 cm^3 under adiabatic conditions. Find the new pressure and temperature of the gas if $\gamma = 1{\cdot}4$.

9. Explain why the value of the specific heat capacity of a gas depends upon the conditions under which the gas is heated. Derive an expression for the difference between the *principal* specific heat capacities of a gas, stating any assumptions made. What is the importance of the *ratio* of these specific heat capacities of a gas?

In a constant-flow experiment it is found that 144 joules are required to raise the temperature of 2·00 g of hydrogen through 5·00 °C when it passes through the apparatus at constant pressure. Given that the density of hydrogen at s.t.p. is $9{\cdot}0 \times 10^{-2}$ kg m^{-3}, find the specific heat capacity of hydrogen *at constant volume*. (*L.*)

10. Derive an expression for the difference between the specific heat capacities of an ideal gas.

Calculate this difference for an ideal gas which has a density of 0·089 kg m^{-3} at s.t.p. (*N.*)

11. Simple kinetic theory leads to the relation $p = \frac{1}{3}\rho\overline{c^2}$ between the pressure p exerted by an ideal gas, its density ρ and the mean square velocity $\overline{c^2}$ of its molecules. What information does this expression, when correlated with the perfect gas equation $pV = RT$, give about the internal energy of an ideal monatomic gas? Explain why it is the mean square velocity of the molecules rather than the square of their mean speed which appears in the relation.

From these two equations derive (*a*) an expression for the difference between the principal specific heat capacities of such a gas, and (*b*) a numerical value for γ, the ratio of its principal specific heat capacities.

Outline one method by which γ has been determined for a real gas. (*O. and C.*)

12. Explain why a gas cools when it is subjected to a reversible adiabatic expansion.

An experimenter plans to measure the ratio of the principal specific heat capacities of a gas by compressing the gas at room temperature in a flask and measuring the temperature change of the gas in the flask when some of the gas is allowed to leak out until all the pressure in the flask is atmospheric. Explain the principle of his proposed method, and describe the experimental difficulties he might encounter. How, without changing the basis of the method, might these difficulties be overcome?

Calculate the final temperature of a perfect gas initially at 300 K when its volume is doubled adiabatically and reversibly if the ratio of its principal specific heats is 5/3. (*O. and C.*)

13. Explain the meaning of the terms *isothermal, adiabatic.* What is the importance of the ratio of the principal specific heats of an ideal gas?

Air initially at 27 °C and at 75 cm of mercury pressure is compressed isothermally until its volume is halved. It is then expanded adiabatically until its original volume is recovered. Assuming the changes to be reversible, find the final pressure and temperature. [Take the ratio of the specific heat capacities of air as 1·40.] (*L.*)

14. Explain why the specific heat capacity of a gas is greater if it is allowed to expand while being heated than if the volume is kept constant. Discuss whether it is possible for the specific heat capacity of a gas to be zero.

When 1 g of water at 100 °C is converted into steam at the same temperature, 2264 J must be supplied. How much of this energy is used in forcing back the atmosphere? Explain what happens to the remainder of the energy. [1 g of water at 100 °C occupies 1 cm^3. 1 g of steam at 100 °C and 76 cm of mercury occupies 1601 cm^3. Density of mercury = 13 600 kg m^{-3}.] (*C.*)

15. Describe an experiment to determine the specific heat capacity of water, at about 15 °C, deriving from first principles any equations used.

Deduce an expression for the difference between the specific heat capacities of an ideal gas. If the specific heat capacity of air at constant pressure is 1·013 kJ kg^{-1} K^{-1} and the density at s.t.p. is 1·29 kg m^{-3}, estimate a value for the specific heat capacity of air at constant volume. [Assume the density of mercury at 0 °C to be 13 600 kg m^{-3}.] (*L.*)

16. Derive, from first principles, an expression for the work done in compressing an ideal gas isothermally.

Estimate the work done in inflating a bicycle tyre. Provide your own approximate data and work in metric units. (*N.*)

17. A litre of air, initially at 20 °C and at 760 mm of mercury pressure, is heated at constant pressure until its volume is doubled. Find (*a*) the final temperature, (*b*) the external work done by the air in expanding, (*c*) the quantity of heat supplied.

[Assume that the density of air at s.t.p. is 1·293 kg m^{-3} and that the specific heat of air at constant volume is 0·714 kJ kg^{-1} K^{-1}.] (*L.*)

18. What is meant by the terms *adiabatic* and *isothermal*? Show that the pressure p and volume V of a given mass of a perfect gas are related by the expression $pV^{\gamma}=$ constant for reversible adiabatic changes, and explain the physical significance of γ.

A mass of air at 20 °C is compressed adiabatically and reversibly to double its initial pressure. What is the final temperature? [$\gamma = 1{\cdot}41$.] (*C.*)

19. Explain why, when quoting the specific heat capacity of a gas, it is necessary to specify the conditions under which the change of temperature occurs. What conditions are normally specified?

A vessel of capacity 10 litres contains 130 g of a gas at 20 °C and 10 atmospheres pressure. 8000 joules of heat energy are suddenly released in the gas and raise the pressure to 14 atmospheres. Assuming no loss of heat to the vessel, and ideal gas behaviour, calculate the specific heat of the gas under these conditions.

In a second experiment the same mass of gas, under the same initial conditions, is heated through the same rise in temperature while it is allowed to expand slowly so that the pressure remains constant. What fraction of the heat energy supplied in this case is used in doing external work? Take 1 atmosphere = 10^5 newton $metre^{-2}$. (*O. and C.*)

20. Define *heat capacity* and *specific heat capacity*.

Describe an experiment to determine the specific heat capacity of a gas *either* at constant volume *or* at constant pressure. Point out likely sources of error and indicate how they may be minimized.

Explain why it is necessary to specify the condition of constant pressure or constant volume. (*L.*)

6
Vapours. Real Gases. Liquefaction

In this chapter we deal with vapours and real gases, the laws each obeys, and the experiments which led to the liquefaction of gases.

Properties of Vapours

Saturated vapours

Consider a vertical tube T about a metre long, which contains mercury and has a vacuum in the space at the top of the tube (fig. 6.1). When a small amount of water is introduced at B by a bent pipette, the water rises to the top of the mercury, because it is less dense, and then evaporates in the empty space. The water-vapour formed exerts a pressure, and the mercury level is depressed by a height equal to the vapour pressure in millimetres of mercury. As more water is introduced at B and changes to vapour in the space above A, the level of the mercury is depressed further, but this effect ceases as soon as water is observed in the liquid state on the top of the mercury column. The water-vapour is now said to be *saturated* (fig. 6.1). The introduction of more water into T has no effect on the water-vapour pressure h, which is known as the *saturation vapour pressure* (s.v.p.) of the water at the temperature of the surroundings. If T is raised a little out of the trough of mercury, so that the saturated vapour expands, the height of the mercury column above the level of the mercury in the trough remains the same, showing that the saturation vapour pressure is unaltered. If T is lowered into the trough so that the volume of the saturated vapour decreases, the mercury height remains unchanged. Thus experiment shows that the saturation vapour pressure is independent of the volume occupied by the vapour.

Differences between unsaturated and saturated vapours

Before water is formed at the top of the mercury column in T, the water-vapour in the space there is said to be *unsaturated vapour* (fig. 6.1). When the volume is increased, observation shows that the

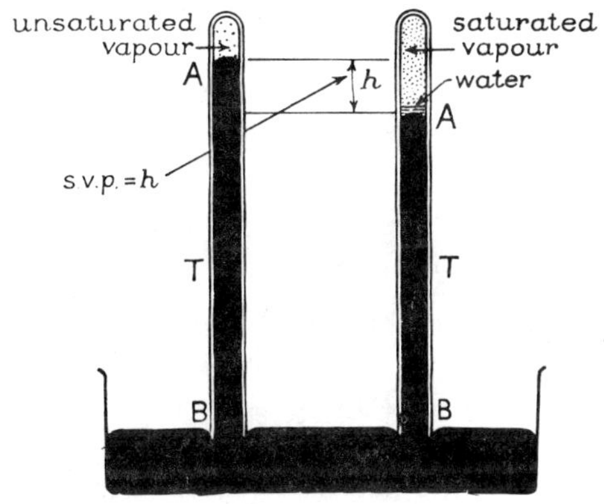

Fig. 6.1 Unsaturated and saturated vapours (h = level below mercury with vacuum present)

pressure of the vapour is decreased. Experiment shows that Boyle's law is approximately obeyed when the pressure values are not near the saturation vapour pressure value for the given temperature. Thus unsaturated and saturated vapours of the same substance have different properties, and the following summary should assist the student to distinguish between them:

(1) A *saturated vapour* is one which is in equilibrium with its own liquid. The pressure exerted by a saturated vapour is always constant for a given temperature of the liquid, even though the vapour undergoes changes in volume. Experiment shows that the saturation vapour pressure of a liquid rises rapidly with its temperature (p. 128).

(2) An *unsaturated vapour* at a given temperature is one which exerts less than the saturation vapour pressure for that temperature, and is obtained when no liquid is in contact with it in a closed space. The pressure of a given mass of unsaturated vapour depends on its volume as well as its temperature, and experiment shows that it obeys Boyle's law approximately.

Volume and temperature changes

Figure 6.2 (i) illustrates the variation of the vapour pressure in a closed space when the volume is increased. At first, with liquid present, the pressure is constant at the saturation value along AB.

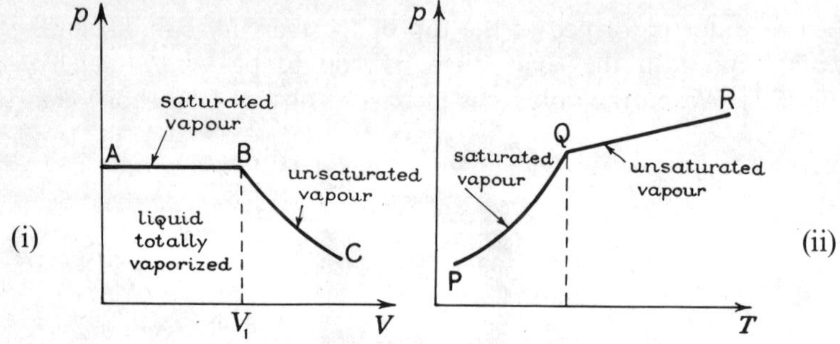

Fig. 6.2 Pressure-volume and pressure-temperature changes

When the volume V_1 is reached, however, all the liquid evaporates and the unsaturated vapour now obeys Boyle's law approximately. Thus the pressure decreases along the curve BC as the volume increases.

Suppose the volume in the closed space is kept constant and the temperature is raised. The saturation vapour pressure rises sharply with temperature rise along PQ (fig. 6.2 (ii)). When the vapour becomes unsaturated, however, it obeys Charles' law fairly well and the pressure p follows a straight line QR since $p \propto T$. The rise of p with temperature is now much less than before, as shown. The line RQ, if produced, would pass through the absolute zero (−273 K).

Kinetic theory explanation of unsaturated and saturated vapours

When a small quantity of liquid at a given temperature is introduced into a closed space at the top of a barometer tube, all the liquid evaporates and the vapour exerts a pressure (p. 123). On the kinetic theory, this phenomenon is explained by assuming that, to start with, more molecules per second leave the liquid and become gaseous in the closed space than return to the liquid from the closed space. This process goes on until no more liquid remains, and an *unsaturated vapour* is obtained.

As more liquid is introduced into the closed space, however, a point is reached when liquid remains there, and the vapour is now in contact with its own liquid. The pressure exerted by the vapour is then constant no matter how much liquid is introduced into the closed space. We therefore speak of the "saturation vapour pressure" of the liquid at the given temperature (p. 123). The density of the vapour has increased to such a value that the same number of molecules of the gas now return to the liquid per second from the closed space as escape from the liquid in that time. A state of *dynamic equilibrium* exists in this case between the molecules which constitute the vapour and those which constitute the liquid. The introduction of further liquid into the tube causes no more molecules per second to leave the liquid, because the speed of the molecules depends only on the temperature of the liquid. Dynamic equilibrium is thus maintained with the same average number of molecules of vapour above the liquid. The vapour is therefore saturated, i.e. it has reached the maximum density possible at the given temperature, and has the maximum possible pressure. If the volume of the space above the liquid is decreased, more molecules of vapour enter the liquid per second than previously; but as the number of molecules leaving the liquid per second remains constant, some of the vapour condenses until dynamic equilibrium is again restored. Conversely, if the volume of the space above the liquid is increased, the density of the vapour again reaches its maximum value by evaporation of some of the liquid, and dynamic equilibrium is once more obtained. Thus the saturation vapour pressure remains constant at a given temperature of the liquid.

When the temperature is increased the speed of the molecules inside the liquid is increased, and the average number of molecules escaping into the space above the liquid hence increases. Dynamic equilibrium again exists if some liquid is present at this higher temperature, but the saturation vapour pressure is higher as the density of the vapour molecules in the space above the liquid is now greater.

Boiling and evaporation

When water is heated in an open vessel its temperature increases, and after a time, observation shows that some bubbles appear inside the liquid, rise to the surface, and then burst and give off vapour. If the external pressure is one atmosphere (760 mm of mercury), and the temperature of the water reaches 100 °C, the whole of the surface of the liquid is filled with bursting bubbles and steam is obtained. *Boiling* is now said to take place. The temperature of the water remains constant at 100 °C as long as the water changes to vapour.

It is possible for vapour to be given off from water at temperatures well below 100 °C. Thus a puddle of water disappears gradually, even though the average temperature of the water may be about 10 °C. This process, which is essentially different from boiling, is known as *evaporation*. Although the temperature of the liquid may be constant, so that the average energy of the liquid molecules is constant, the *individual* molecules are moving with different speeds. When a molecule near the liquid surface has a sufficient amount of kinetic energy, it is able to overcome the attraction of the molecules of the liquid round it and escape through the surface of the liquid to the outside, where it constitutes a molecule of vapour. This is the process which takes place continuously when liquid such as a pool of water evaporates in an open space; the molecules which escape from the liquid are carried away by the wind, which thus assists further evaporation. Since the faster molecules escape, the average kinetic energy of those left in the liquid decreases. Thus the liquid cools when evaporation occurs.

Boiling and condensation

When a bubble of radius r is formed in a liquid, the pressure p inside it is the saturation vapour pressure of the liquid. This is greater than the pressure outside by $2\gamma/r$, where γ is the surface tension of the liquid. Thus if B is the atmospheric pressure, h is the depth of the bubble in the liquid, and ρ is the liquid density,

$$p = \frac{2\gamma}{r} + B + h\rho g \quad \text{. . . . . (i)}$$

If r is large, p is only very slightly greater than B, from (i). On the other hand, if r is very small, of the order of 10^{-4} mm for example, calculation from (i) shows that p will then be much greater than B and of the order of several atmospheres for water. To reach such a high value of p, the saturation vapour pressure, the liquid must be heated to a high temperature. Thus water must be heated to about 120 °C to have a saturation vapour pressure of about 2 atmospheres, and to about 142 °C for 4 atmospheres pressure.

When water is boiled in a kettle, the rough parts of the surface act as centres or nuclei of large radius on which bubbles can form. As just explained, the saturation vapour pressure is then only slightly greater than atmospheric pressure. Thus when boiling occurs the bubbles burst quickly. Dissolved air-bubbles also act as nuclei of large radius.

If, however, water is boiled in a vessel with very smooth surfaces and is air-free, there are no nuclei of large radius on which the bubbles can form. In this case the temperature will rise well above 100 °C before a bubble begins to form on a tiny irregularity of the surface. It then expands very rapidly owing to the high pressure inside, and the liquid is forced violently against the sides of the vessel. This produces the "bumping" heard. It can be overcome by placing small pieces of unglazed porcelain, or porous material, in the vessel. These provide large nuclei on which the bubbles can form, and boiling then proceeds quietly.

In a similar way, drops cannot form in saturated vapours unless there are dust particles or other centres of appreciable radius on which they can begin to form. A perfectly dust-free space can thus contain more water-vapour than normal, and is then said to be *supersaturated.* Drops form readily round ions, however, which may be produced in the dust-free space. This is due to the outward stress on the surface of a charged drop, which tends to neutralize the inward pressure due to surface tension. It thus reduces the effect of the term $2\gamma/r$ in equation (i) for small values of r. The drops which form round ions in supersaturated vapour are used in cloud chambers to show the paths of ionizing particles or radiation.

Effect of pressure on boiling-point

Observation shows that water boils at a much lower temperature than 100 °C at the top of a mountain; at 5000 metres, for example, water boils at 83 °C. In the past, mountaineers deduced the height they had ascended by finding the boiling-point of water at this height, and then consulting tables showing the variation of the boiling-point with height. It thus appears that reduced external pressure on a liquid leads to a reduction in the boiling-point.

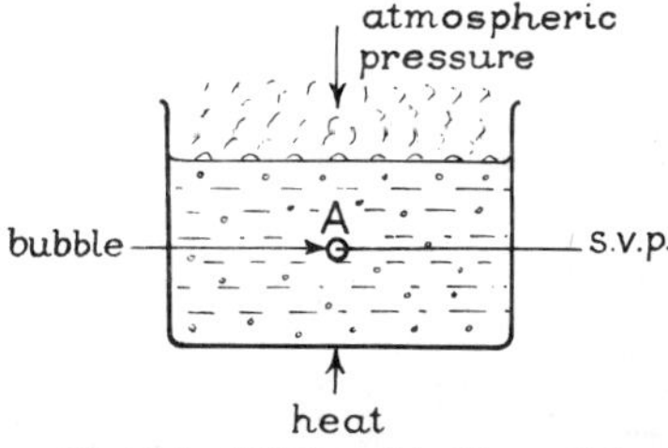

Fig. 6.3 S.V.P. and boiling-point

We can explain this effect by supposing that some water is heated in a vessel and a bubble A is formed in the liquid (fig. 6.3). The external pressure on the bubble is equal to the atmospheric pressure plus the pressure due to the small head of water above A. Inside the bubble the pressure is equal to the saturation vapour pressure of water at the temperature of the liquid, assuming there is only water-vapour inside A, and if the saturation vapour pressure is less than the external pressure the bubble cannot grow. As the water is heated, the saturation vapour pressure, which increases with temperature, reaches a value when it just exceeds the pressure outside the bubble. The bubble then grows and rises quickly to the surface of the water, where it bursts. Vapour is now given off, and the liquid is said to be boiling. Thus *boiling occurs when the temperature of the liquid is such that its saturation vapour pressure is equal to the external atmospheric pressure*. If the atmospheric pressure decreases, the liquid boils at a lower temperature because the saturation vapour pressure necessary for boiling is now less. Similarly, increased pressure raises the boiling-point.

The boiling-point of a liquid is influenced by impurities in it, as well as by the external pressure. The boiling-point of water *increases* when salt is added to it, for example, and the boiling-points of other liquids are also increased when solid impurities are added. There is no simple relation between the elevation of the boiling-point of a liquid and the impurities dissolved in it, but when the mass m of impurity is small, the elevation of the boiling-point is approximately proportional to m.

Measurement of saturation vapour pressure (s.v.p.)

We have now to consider methods of measuring the s.v.p. of water, which vary according to the temperature. The magnitude of the s.v.p. of water is required in the design of steam turbines, in meteorology, and in humidifiers, for example.

(*a*) *From* 0–50 °C. For this range of temperature, the method illustrated in fig. 6.4 can be used. Two long vertical tubes, M, N, filled with mercury, are placed in a trough, and the mercury levels in both tubes are originally at the same level corresponding to A, with a vacuum above each. Water is now continually introduced into M by means of a bent pipette, and the level of mercury is depressed until some water W remains on the mercury at B. The difference in height h between A and B is then noted, and corresponds to the s.v.p. at the temperature of the surroundings after correction is made for the small amount of water W and the surface tension effect. A liquid bath D surrounds the upper part of both tubes, and

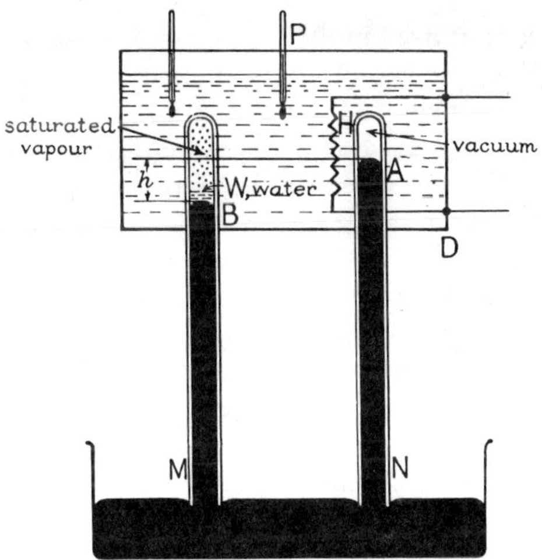

Fig. 6.4 s.v.p. of water by static method

an electrical heating coil H in the liquid enables the temperature to be varied. The temperature is read from thermometers P. D has plane glass windows to enable accurate observations to be made on the mercury levels by a travelling microscope. In this experiment, it should be noted, the level of A acts only as a reference line and remains constant.

(*b*) *Above* 50 °C. We have already seen that a liquid boils when its saturation vapour pressure is just equal to the pressure of the atmosphere outside the liquid (p. 128). Thus if a liquid boils at t °C the s.v.p. at this temperature is the same as the external pressure.

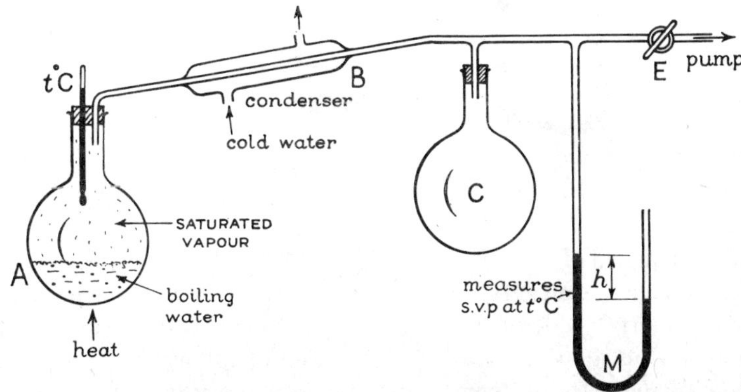

Fig. 6.5 s.v.p. of water by dynamic method

This principle is utilized in the measurement of saturation vapour pressures at different temperatures. The liquid A is contained in a vessel connected through a sloping tube B to a bulb C and a manometer M (fig. 6.5). B is surrounded by water flowing through a condenser, so that the vapour formed is condensed and the liquid obtained is used again, while the bulb C helps to prevent large fluctuations of pressure in the apparatus. The pressure of the air in the system is first fixed at some value by connecting a pump at E and then closing this tap, and the liquid A is now heated. After a time the liquid boils at a temperature indicated by the thermometer above A, and the corresponding pressure obtained from the manometer M is the s.v.p. of the liquid at this temperature. By varying the pressure, the variation of s.v.p. with temperature is obtained. A strong metal vessel and tubes should be used for measurement of s.v.p. above atmospheric pressure, in which case air must be pumped into the apparatus.

(*c*) *Below* 0 °C, the s.v.p. of a liquid can be found by means of the apparatus shown in fig. 6.6. A bulb S is connected to a manometer M, and S is surrounded by a freezing mixture whose temperature is measured by means of a thermometer T. The space between S and M

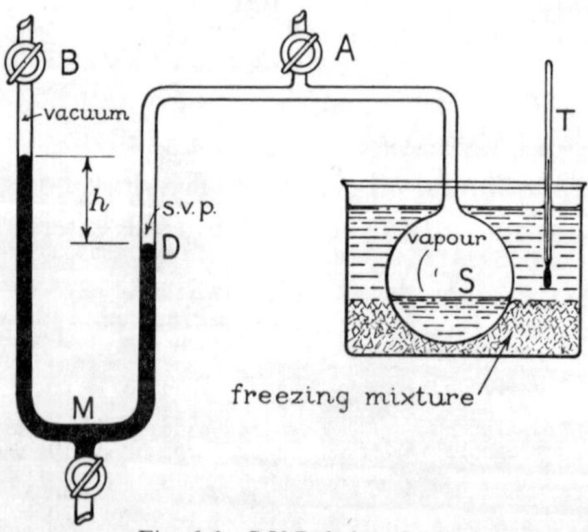

Fig. 6.6 S.V.P. below 0 °C

is first exhausted of air, and some vapour of the substance investigated is introduced through A until liquid is formed in S. The mercury level on the right side of M, which had been depressed as soon as the vapour was introduced, now remains constant, and if the space below B is a vacuum, the difference in levels h of the mercury column is equal to the s.v.p. at the temperature registered on T.

Although the vapour at D is not at the temperature of the freezing mixture, the pressure measured by M does correspond to this temperature. To prove this, suppose that the pressure of the vapour at D is higher than that above the liquid in S. Vapour then flows from D to S to equalize the pressure, and some condenses to liquid. Similarly, if the vapour pressure at D is less than that above the liquid in S, some of the liquid evaporates to equalize the pressure. Thus the vapour pressure in D is always equal to the s.v.p. of the liquid in S.

Variation of s.v.p. of liquid with temperature

In general, the s.v.p. of a liquid increases as the temperature increases, but there is no simple relationship between the two quantities. Figure 6.7 illustrates roughly the variation of the s.v.p. of water in mm of

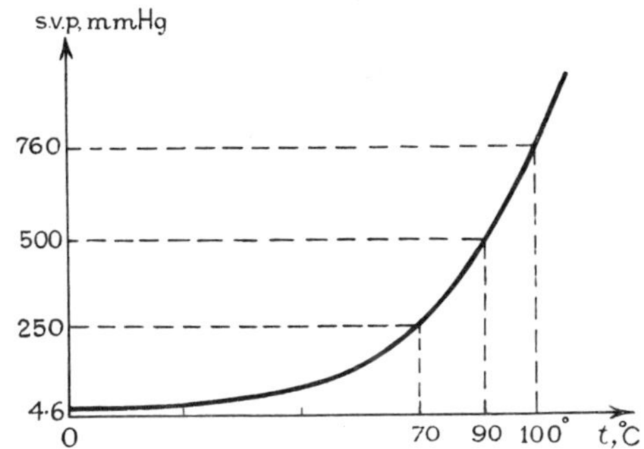

Fig. 6.7 Variation of s.v.p. with temperature (not to scale)

mercury with temperature in °C; at 100 °C the s.v.p. is 760 mm mercury, at 0 °C it is 4·6 mm mercury.

Dalton's Law of Partial Pressures

Dalton put forward a law, known by his name, relating to the pressure of gases in a mixture. The law states:

In a mixture of gases, each gas exerts a pressure which is the same as if it were alone and occupying the volume of the mixture.

The pressure of a particular gas in the mixture is known as a "partial" pressure because other gases are present, and the *total* pressure of the gases is the sum of all the partial pressures. Thus if the total pressure is a mixture of two gases is 740 mm of mercury and the pressure of one gas is 18 mm, the pressure of the other gas is 740 – 18 = 722 mm. The volume occupied by either gas is the volume of the mixture, and we can understand this to be the case because the molecules of either gas move about in the whole of the space occupied by the mixture.

Mixture of gas and saturated water-vapour

Suppose that some water and air are present in the closed space at the top of a column of mercury, the temperature being 10 °C. If the s.v.p. of water at 10 °C is 8 mm of mercury, and the total pressure of the water-vapour and air is 700 mm mercury, the pressure of the air $= 700 - 8 = 692$ mm, from Dalton's law. Suppose that the volume of the mixture is 30 cm^3, and that the volume is reduced to 25 cm^3, the temperature remaining constant. Then if p is the new total pressure of the mixture, the pressure of the air is now $(p-8)$ mm, as the s.v.p. of water depends only on its temperature. Further, the volume originally occupied by the air was 30 cm^3, and 25 cm^3 when the volume is altered.

To find the magnitude of p, we apply Boyle's law to the given mass of *air* in the mixture, since the temperature is unchanged. As pressure × volume is constant,

$$\therefore (p-8)25 = 692 \times 30$$

$$\therefore p = 838 \text{ mm}$$

It will be noted that any of the gas laws, such as Boyle's or Charles' law, for example, can be applied to a particular gas in a mixture, provided that the mass of this gas remains constant. The choice of the gas to which a gas law can be applied is limited to air in the case of an air and water-vapour mixture, as some of the water-vapour may condense, or some of the water present may vaporize, when the pressure, volume, or temperature is altered. This results in a change in the mass of water-vapour present in the mixture, so that the gas laws, which apply to a *constant* mass of gas, cannot be applied to water-vapour in contact with water.

Variation of total pressure in a mixture of air and water-vapour

Figure 6.8 (i) illustrates how the pressure varies with volume in a closed space containing a little water, water-vapour and air. As the volume is increased the water-vapour pressure remains constant along AB so long as it is saturated vapour; when it becomes unsaturated vapour it obeys Boyle's law and follows the curve BC (see p. 133). Air pressure, however, follows a Boyle's law variation continuously

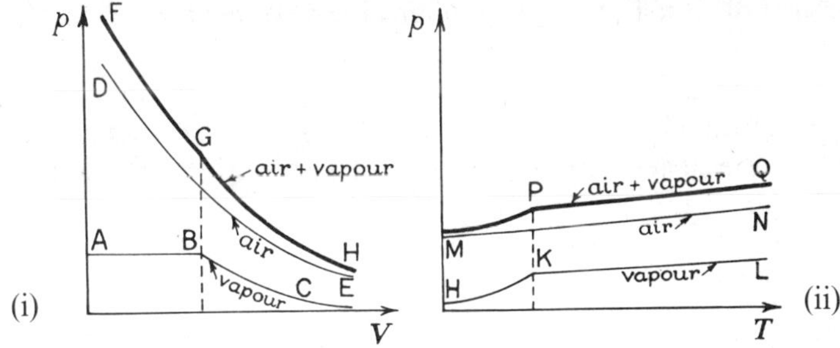

Fig. 6.8 (i) Volume; (ii) temperature change on air-vapour mixture

along DE. The total pressure variation is obtained by adding the two curves and is shown roughly by FH.

Suppose now that the volume of the closed space is kept constant and that the temperature is raised when water is initially present. The rise in s.v.p. at first follows the curve HK; when the water-vapour becomes unsaturated the pressure varies along a straight line KL like a gas (fig. 6.8 (ii)). The air pressure, however, follows the parallel straight line MN throughout. The total pressure is the sum of the two curves and is shown roughly by MPQ.

Measurement of s.v.p. using Dalton's law

Dalton's law can be used as the basis of a simple method of determining the variation of the s.v.p. of water with temperature, when the s.v.p. at one temperature is known. A pellet of water W is contained in a uniform capillary tube T, open at one end so that a mixture of air and saturated water-vapour is obtained in a closed space below the pellet (fig. 6.9). The tube is placed inside a liquid bath, and the

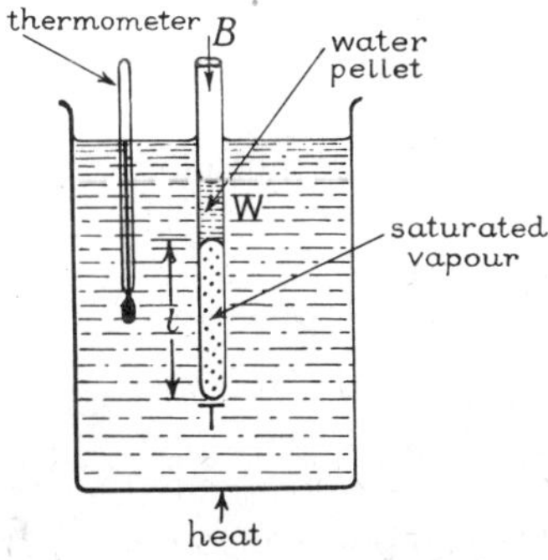

Fig. 6.9 s.v.p. of water by capillary tube method

length of the capillary tube containing the mixture of gases is measured as the temperature is varied.

Suppose B is the atmospheric pressure, a the uniform cross-sectional area of the tube, l and l_1 the lengths of the tube containing the mixture at temperatures t, t_1 °C respectively. Then the pressure of the air in the closed space at t °C is $B-p$, where p is the s.v.p. of water at t °C, as the total pressure of the gases in the closed space is always equal to the external (atmospheric) pressure, if we neglect the weight of W. Similarly, the pressure of the same mass of air at t_1 °C is $B-p_1$, where p_1 is the s.v.p. of water at t_1 °C. The volume of the air at t_1 °C $= l_1a$, and the volume at t °C $= la$. Hence, since pV/T is constant for a given mass of air, we have

$$\frac{(B-p_1)l_1a}{273+t_1} = \frac{(B-p)la}{273+t}$$

$$\therefore \frac{(B-p_1)l_1}{273+t_1} = \frac{(B-p)l}{273+t}$$

Knowing B, l_1, l, t_1, t and the s.v.p. p at temperature t °C, the s.v.p. p_1 at the temperature t_1 °C can be calculated from this relation. The s.v.p. at other temperatures is calculated from similar relations.

Relative humidity

On fine days the air feels clear and dry; on "muggy" days the air feels damp and oppressive. Nevertheless there may be much more water-vapour in the air in the former case than on a muggy day, so that our judgment of the degree of humidity of the air cannot be based only on the actual amount of water-vapour in the air. The important point is: *How near to saturation* is the water-vapour in the air? If it is very close to saturation we cannot perspire freely, as the water-vapour already in the air is close to the amount required to saturate the same volume of air. Conversely, we can perspire freely if the water-vapour is not close to the saturation value, and we are correspondingly comfortable. Scientists define the *relative humidity* (r.h.) of the air as the ratio given by:

$$\text{relative humidity} = \frac{m}{M} \times 100\%$$

where m is the mass of water-vapour in a given volume of air, and M is the mass of water-vapour required to *saturate* the same volume at the same temperature. A high value of relative humidity means that the air is nearly saturated; a low value means that the air is far from being saturated.

Since water-vapour obeys the gas laws fairly well, the mass of water-vapour in a given volume of air is proportional to its partial pressure. Scientists therefore take as a working formula for relative humidity the ratio defined by:

$$\text{relative humidity} = \frac{p}{P} \times 100\,\%$$

where p is the actual vapour pressure in a volume of air and P is the saturation vapour pressure in the same volume of air.

Dew-point

In the early morning, when the temperature is low, dew may be observed on grass, showing that the air near the grass has become saturated with water-vapour. The temperature at which the air becomes saturated with the water-vapour present in it is known as the *dew-point*, and the latter can be determined by progressively cooling a bright metal surface in the air. At some point the surface becomes misty, showing that water has condensed on it, and the dew-point is the corresponding temperature of the cooled metal surface.

The dew-point is utilized in one of the standard methods of measuring relative humidity. Suppose that the temperature of the air in a room is 15 °C and that dew is deposited on a metal surface, as it is cooled, when its temperature is 10·8 °C. Then the pressure of the water-vapour in the room at 15 °C is equal to the s.v.p. of water at 10·8 °C. It follows from the definition of relative humidity given above that generally

$$\text{relative humidity} = \frac{\text{s.v.p. of water at the dew-point}}{\text{s.v.p. of water at the air temperature}} \times 100\,\%$$

Thus one method of measuring relative humidity consists basically of (*a*) finding the dew-point, (*b*) then referring to tables giving the s.v.p. values for water at different temperatures, and (*c*) using the relation above. Details of measuring relative humidity can be found in more elementary works, such as *Fundamentals of Physics* (Chatto & Windus) by the author.

Examples

1. A vessel contains a mixture of air and water-vapour in contact with excess of the liquid. How will the pressure in the vessel change, (*a*) if the volume is changed at constant temperature, (*b*) if the temperature changes at constant volume?

A closed vessel contains a mixture of air and water-vapour in contact with excess of water. The pressures in the vessel at 27 °C and 60 °C are respectively 777 mm and 981 mm of mercury. If the vapour of water at 27 °C is 27 mm of mercury, what is the vapour pressure at 60 °C? (*O. and C.*)

First part.—(*a*) If the volume is reduced at constant temperature, the pressure of the air increases, by Boyle's law. The s.v.p. of the water remains unchanged, however, and the total pressure of the mixture thus increases. If the volume is increased, similar reasoning shows that the total pressure diminishes. (*b*) If the temperature is increased at constant volume, the pressure of the air and the s.v.p. of water both increase; hence the total pressure increases. Similar reasoning shows that the total pressure diminishes if the temperature diminishes. See p. 131.

Second part.—The pressure of the air at 27 °C = 777 − 27 = 750 mm
The pressure of the air at 60 °C = 981 − p

where p is the s.v.p. of water at 60 °C. Since the pressure of a gas at constant volume is proportional to its absolute temperature (p. 45), we have, for the constant mass of air in the mixture,

$$\frac{981-p}{750} = \frac{273+60}{273+27}$$

$$\therefore 981-p = \frac{333}{300} \times 750$$

$$\therefore p = 149 \text{ mm}$$

2. Explain what is meant by the "dew-point", and describe an accurate method of determining its value. The saturated vapour pressure of water at 11·25 °C is 1 cm of mercury, the density of dry air at 76 cm pressure and 25 °C is 0·001 184 g/cm^3, and the density of water-vapour relative to that of air at the same pressure and temperature is 0·624. Find the weight of a litre of moist air at 25 °C, if the barometer stands at 76 cm and the dew-point is 11·25 °C. (*W.*)

Since the dew-point is 11·25 °C, the s.v.p. of the water = 1 cm of mercury. Hence the pressure of the air at 25 °C = 76 − 1 = 75 cm. The mass of 1000 cm^3 of air at 76 cm pressure and 25 °C is 1·184 g. At 75 cm pressure and 25 °C the volume of this air is 1000 × 76/75 cm^3 from Boyle's law.

∴ mass of 1 litre of air at 75 cm pressure and 25 °C

$$= 1{\cdot}184 \times \frac{1000}{1000 \times 76/75} = 1{\cdot}184 \times \frac{75}{76} = 1{\cdot}169 \text{ g}$$

We have now to find the mass of 1 litre of water-vapour at 1 cm pressure and 25 °C, since the pressure of the water-vapour in the moist air is 1 cm. As the mass of 1 litre of air at 1 cm pressure and 25 °C is 1/75 × 1·169 g, and the density of water-vapour is 0·624 of that of air at the same pressure and temperature, the mass of 1 litre of water-vapour at 1 cm pressure and 25 °C is

$$0{\cdot}624 \times \frac{1}{75} \times 1{\cdot}169 = 0{\cdot}0097 \text{ g}$$

$$\therefore \text{mass of moist air} = 1{\cdot}169 + 0{\cdot}0097 = 1{\cdot}179 \text{ g}$$

Vapour-Gas Relationship. Critical Temperature

Isothermals of carbon dioxide

In 1869 Andrews investigated the relation between the pressure and volume of carbon dioxide gas at constant temperatures. As we see later, his results are of great importance in the liquefaction of gases (p. 138).

Andrews subjected carbon dioxide gas to various pressures, and noted the volumes obtained, keeping the temperature constant each time at different values. The principle of the apparatus he used is illustrated by fig. 6.10 (i). Carbon dioxide (CO_2) is contained in a capillary tube A by a mercury pellet. Next to A is another capillary tube B containing air, which is also kept in the tube by a mercury pellet. The open ends of the tubes lead to the same liquid chamber, and by means of the plungers M, M, which can be screwed into the

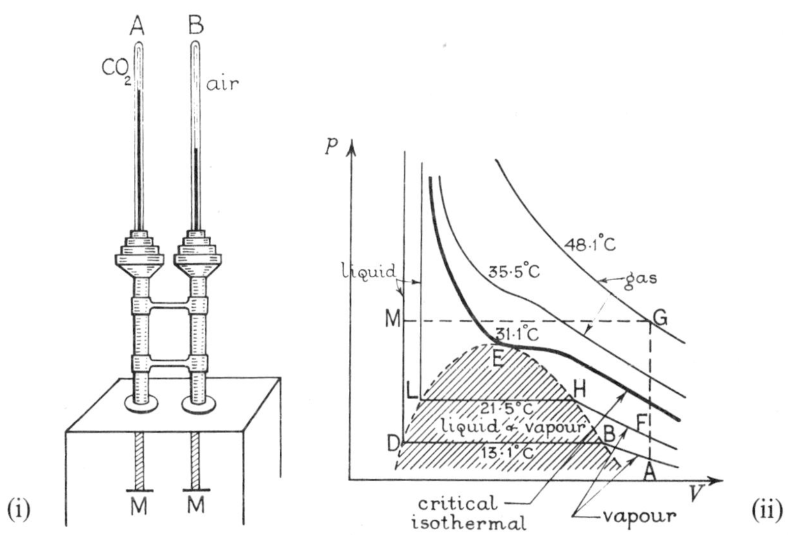

Fig 6.10 (i) Andrews' experiment on CO_2 and (ii) results

chamber, the carbon dioxide and air are subjected to the same pressures. Andrews had previously determined the volume of the air in B at *known* pressures. He was thus able to find the pressure on the carbon dioxide by noting the corresponding volume of the air in B.

The results obtained are illustrated in fig. 6.10 (ii), which represents some of the isothermals, or $p-V$ curves at constant temperature, of carbon dioxide. As the carbon dioxide vapour is compressed from a condition represented by A on the *isothermal of* 13·1 °C, a point is reached at B when liquid carbon dioxide begins to be formed. The vapour is then saturated, and the pressure remains constant while its volume decreases until the point D is reached. The carbon dioxide is now completely liquid, and consequently further pressure decreases the volume only slightly, as shown by the nearly vertical line DM. As shown by the *isothermal of* 21·5 °C, the carbon dioxide exhibits the same changes in state as it did at 13·1 °C, but the volume change on liquefying is less than before since HL is smaller than BD.

As the temperature is increased further, the decrease in volume, corresponding to the horizontal portion of the isothermal, diminishes, and at 31·1 °C the liquid state is barely obtained, as shown by the *isothermal of* 31·3 °C. At the condition represented by E on the isothermal, the meniscus of liquid carbon dioxide will disappear. Above 31·1 °C, as shown by the *isothermal of* 35·5 °C, no liquid is formed, and the "kink" in the curve becomes less. The *isothermal of* 48·1 °C shows that the gas now obeys Boyle's law, as the curve follows the law $pV =$ constant.

Critical temperature and pressure. Continuity of state

The most important deduction from Andrews' experiments is that carbon dioxide gas cannot be liquefied if its temperature is above 31·1 °C, no matter to what degree of pressure the gas is subjected. The temperature of 31·1 °C is accordingly known as the critical temperature of carbon dioxide, and the corresponding isothermal is known as the *critical isothermal.* Every gas has a critical temperature which depends on its nature and, generally, *the critical temperature of a substance may be defined as the temperature above which it is impossible to liquefy it.* The pressure of a substance at its critical temperature is known as its critical pressure, and the volume per unit mass of a substance at its critical temperature and pressure is known as its *critical volume*. The critical pressure of carbon dioxide, corresponding to the point E in fig. 6.10 (ii), is about 73 atmospheres. Andrews' experiment has thus an important connection with the methods of liquefying gases. The critical temperature of nitrogen is −147 °C, for example, and the temperature of the gas must hence be lower than −147 °C in order to liquefy it.

Andrews' experiments also illustrate the *continuity* which exists between the liquid and gaseous states of a substance. Above a temperature of 31·1 °C, carbon dioxide exists only as a gas; below this temperature it can exist as a liquid or a vapour, or as a mixture of both states. The mixture of vapour and liquid corresponds to the shaded area in fig. 6.10 (ii). The change from the vapour to the liquid state, however, can be made without vapour and liquid being both present at some intermediate stage. Thus, suppose the volume of vapour in the condition A is kept constant, and the pressure is increased by heating until the point G is reached (fig. 6.10(ii)); the liquid state can then be achieved at once by keeping the pressure constant and decreasing the volume by cooling until M is reached. These arguments show that *one* equation should fit *all* isothermals of a substance, such as those in fig. 6.10 (ii).

Van der Waals' equation

The equation $pV = RT$ is the relation between the pressure, volume, and absolute temperature of unit mass of an *ideal* gas, i.e. one in which the attraction between the molecules, the volume occupied by the molecules, and the time of collision are all assumed to be zero. These are some of the assumptions which form the basis of the simple kinetic theory of gases (p. 56). In 1877 Van der Waals proposed a gas equation which applies better to real gases than the equation $pV = RT$. We shall consider briefly how this equation is derived.

If the molecules in unit mass of a gas occupy a volume represented by b, a constant for this mass of gas, the "effective" or *co-volume* of the gas is $(V-b)$. In place of V in the perfect gas equation $pV=RT$, we must hence write $(V-b)$.

To take into account the effect of the attraction between the molecules of a real gas, consider the molecules striking the face of a piston containing the gas. The molecules inside the gas attract these molecules with a force directed inwards, and hence the pressure p on the piston is less than the pressure would be in the absence of inter-molecular attraction. In the latter case, the pressure would be $p+p_1$, where p_1 is due to the attractive force of the molecules. Now if we imagine the same number of molecules of the gas to occupy a larger volume, i.e. the density of the gas decreases, it can be seen that the force of attraction between the molecules decreases because they are then farther apart. If the density of the gas increases, the force of attraction increases. Since the force is proportional to the *product* of the number of molecules striking the piston and the number inside the gas, it follows that the force is proportional to the *square* of the density ρ of the gas. Thus $p_1 \propto \rho^2$. But the density of a given mass of gas is inversely proportional to its volume V. Hence p_1

varies as $1/V^2$, or $p_1 = a/V^2$, where a is a constant for a given mass of gas.

The equation for unit mass of a real gas can thus be written as

$$\left(p + \frac{a}{V^2}\right)(V - b) = RT \quad . \quad . \quad . \quad . \quad . \quad (1)$$

This equation was first derived by Van der Waals and is associated with his name.

Shape of carbon dioxide isothermals

Van der Waals' equation is a cubic equation in V for a given value of p and T. This can be seen by multiplying together the brackets in equation (1) and then multiplying the whole by V^2. The shape of the p-V curve of van der Waals' equation is thus similar in shape to the isothermals above the critical temperature for carbon dioxide gas, obtained in Andrews' experiments (see fig. 6.10).

Van der Waals' equation can be used to calculate theoretically the values of the critical temperature, pressure and volume. Thus for the isothermal corresponding to the critical temperature, the curve becomes flat or horizontal at the critical point (see fig. 6.10). Here, then, the three roots of the cubic equation coincide, that is, the equation now has the form $(V - V_c)^3 = 0$, where V_c is the critical volume. On expansion and comparing the coefficients, V_c can be found from the known values of a and b. This leads to calculated values for the critical pressure and temperature, p_c and T_c. The

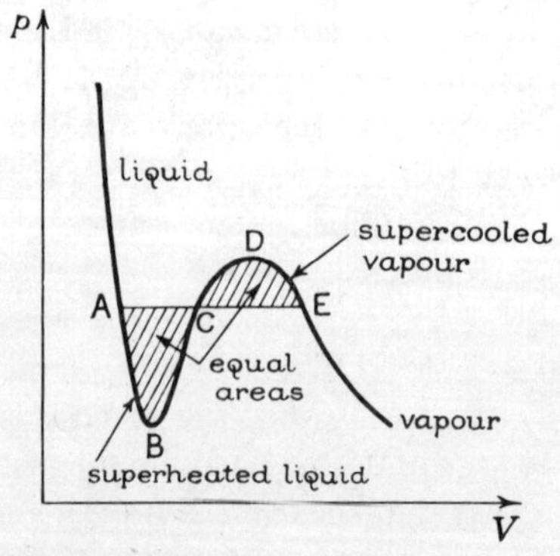

Fig. 6.11 Van der Waals equation—continuity of state

calculated values agree only moderately well with the experimental values, and this is the case for many gases.

Below the critical temperature, when the straight horizontal vapour-liquid stages appear, a cubic or S-shaped curve can still be obtained *theoretically*. This is done by drawing a curve with equal areas above and below the straight line AE, as shown in fig. 6.11. The part AB of the curve represents a region of superheating of the liquid; the part ED represents a region of supercooling of the vapour. Superheating and supercooling can both be observed under special conditions. The part BCD of the curve can never be observed, however, as it represents an expansion when the pressure is increased. The whole curve shows a "continuity of state"—the substance changes from liquid to vapour along this curve.

Many other gas equations, besides van der Waals' equation, have been proposed for actual gases, but no gas equation has yet been found to represent the behaviour of an actual gas over the *whole range* of its volume, pressure, and temperature.

Distinction between "gas" and "vapour"

We are now in a position to understand the scientific difference between a gas and a vapour. When a substance is below its critical temperature, e.g. carbon dioxide at room temperature, we refer to the substance as "vapour", thereby implying that the liquid form is easily obtainable. Water has a critical temperature of about 370 °C and hence we usually refer to "water vapour". When the substance is usually *above* its critical temperature, however, we refer to it as a "gas" since it cannot be liquefied. The critical temperatures of hydrogen and nitrogen are respectively −241 °C and −147 °C, and hence these substances are normally well above their critical temperatures. Hydrogen and nitrogen are consequently known as "permanent" gases. The terms "vapour" and "gas" are artificial distinctions; above the critical temperature a substance is usually termed a "gas", and below the critical temperature it is termed a "vapour".

Departure from Boyle's law

Boyle's law, $pV = \text{constant}$ at a fixed temperature, is an ideal gas law (p. 46). The apparatus in Andrews' experiment is suitable for investigating how the product pV varies with p for real gases such as carbon dioxide which can be easily liquefied. It is unsuitable for gases such as hydrogen or nitrogen. These so-called permanent gases require much higher pressures, such as several hundred atmospheres or more. The tube containing the gas is therefore placed in a steel vessel, and arrangements are made to equalize the pressure outside and inside by surrounding the vessel with water.

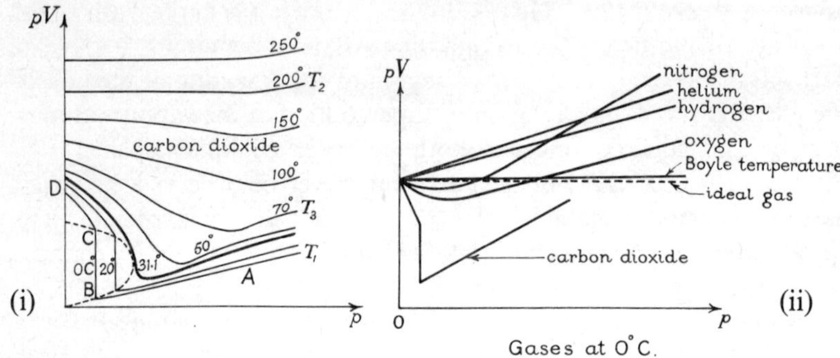

Fig. 6.12 Curves of pV against p

Figure 6.12(i) illustrates the general results for pV against p at different constant temperatures. For temperatures such as T_1 *below the critical temperature*, the variation follows a typical curve ABCD similar to that obtained in Andrews' investigations—at B the gas liquefies and p is constant along the vertical line BC. *Above the critical temperature*, for example at T_3, the value pV decreases to a minimum and then rises. At high temperatures such as T_2, the curve slopes upwards from the beginning and pV increases as p increases from zero. Figure 6.12(ii) compares typical results for different gases at 0 °C.

Virial coefficients. Boyle temperature

For any curve in fig. 6.12(i), pV can be expressed in terms of p by a relation of the form

$$pV = A + Bp + Cp^2 + Dp^3 + \quad . \quad . \quad . \quad . \qquad \text{(i)}$$

The coefficients A, B, C, D are called *virial coefficients.* They are constant at a particular temperature but vary with temperature. Their magnitude diminishes rapidly in the order A, B, C, D.

For an ideal gas, $pV = RT$. Hence, from (i), $A = RT$ when p tends to zero. Thus all gases obey Boyle's law at infinitely low pressures. On this account the gases such as helium in gas thermometers have low pressures and the slight divergences from ideal gas laws are corrected by extrapolating to zero pressure, using the virial coefficients.

At ordinary temperatures B is negative for oxygen, nitrogen, and air, and positive for helium and hydrogen. C is always positive.

The temperature when $B = 0$ is called the *Boyle temperature* of the gas. In this case $pV = A =$ constant to an excellent approximation, since Cp^2 and higher powers are negligible except at very high pressures. The gas then obeys Boyle's law over a wide range of pressures.

Triple point

The three phases or states of matter, solid, liquid and gas, coexist in equilibrium at a temperature called the *triple point* of the substance. Water has a triple point of about 0·01 °C, defined as 273·15 K (see p. 2), and pressure 4·6 mmHg. At this temperature and pressure, ice, water, and water-vapour coexist in equilibrium (fig. 6.13 (i)).

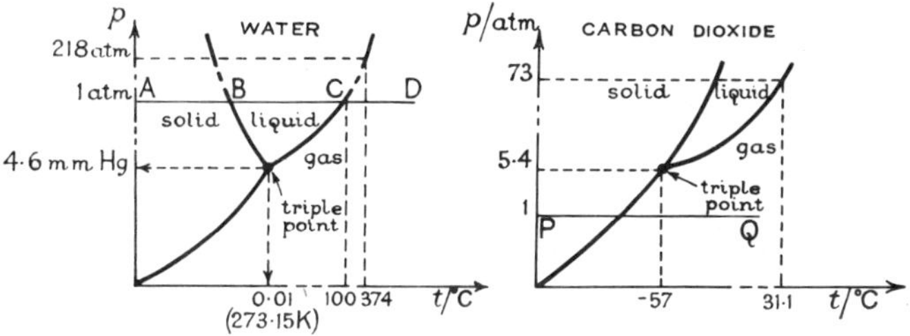

Fig. 6.13 Solid, liquid and gas phases—triple point

The critical temperature and pressure of water are 374 °C and 218 atmospheres respectively. Since 1 atmosphere is a pressure above the triple point of water, ice changes to liquid and then to vapour on warming at normal atmospheric pressure. The phase change occurs along the line ABCD in fig. 6.13 (i).

In contrast, carbon dioxide has a triple point of about −57 °C and a pressure of 5·4 atmospheres (fig. 6.13 (ii)). 1 atmosphere is *below* this pressure. Thus if solid carbon dioxide at a pressure of 1 atmosphere is warmed, it changes along PQ in fig. 6.13 (ii) from the solid to the gaseous state, without an intermediate liquid state. This is called *sublimation*. Iodine is an element which also undergoes sublimation. Solid carbon dioxide, "dry ice", is used as a cooling agent in refrigerators. It is made simply by opening the valve in a cylinder containing the gas at high pressure. The gain in kinetic energy of the released gas is obtained from the internal energy of the gas, since the sudden expansion is substantially an adiabatic change, and the gas is then cooled and passes directly to the solid state.

Liquefaction of Gases

Cascade method

Andrews' experiments showed that a gas can only be liquefied by pressure alone when its temperature is below its critical temperature. Oxygen, which has a low critical temperature, can be liquefied by an arrangement whose principle is illustrated by fig. 6.14. Methyl chloride is liquefied by compression in A, and is then cooled by a water-bath W. The liquid is then allowed to evaporate through a

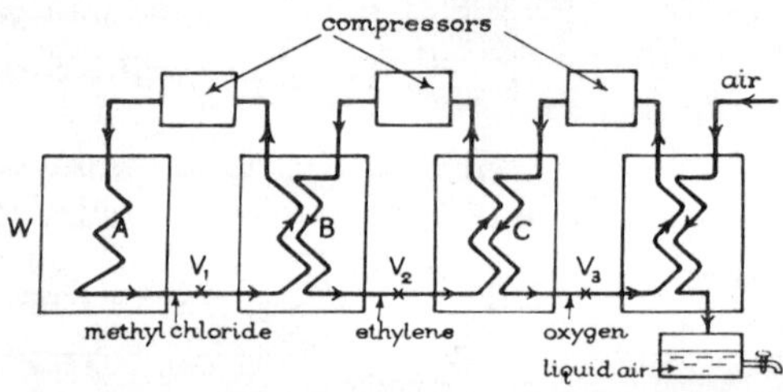

Fig. 6.14 Cascade method of liquefaction

valve V_1 into a vessel under normal atmospheric pressure, and since methyl chloride boils at -24 °C at this pressure, the temperature is sufficiently low to liquefy ethylene, critical temperature 10 °C, by compression. This substance is contained in the pipes B, and when liquid ethylene evaporates under reduced pressure, a temperature of about -150 °C can be obtained. Oxygen, whose critical temperature is -118 °C, can thus be liquefied in the pipes C by compression. By adding another stage, as illustrated by fig. 6.14, Dewar liquefied air by the cascade method in 1898.

Intermolecular forces. Joule-Kelvin effect

If the molecules of a gas have an attraction for each other, the gas will do work against the attractive forces when it expands, since the molecules become more separated. The energy for this *internal work* is provided by the heat in the gas itself if no heat enters or leaves it, and the gas is therefore cooled (see also p. 34).

In Joule's first investigation on intermolecular attraction, he allowed a gas to expand into a vacuum and measured the temperature change (p. 54). However, only the gas initially expanding into the vacuum does no external work; the gas following on compresses that already in the space filled.

In 1853 Joule and Lord Kelvin investigated intermolecular attraction by passing compressed gas at a slow steady rate through a copper spiral tube immersed in a thermostat, and then through a plug of cotton wool or silk P, having a number of fine holes or pores (fig. 6.15). The porous plug produced a large number of slow small

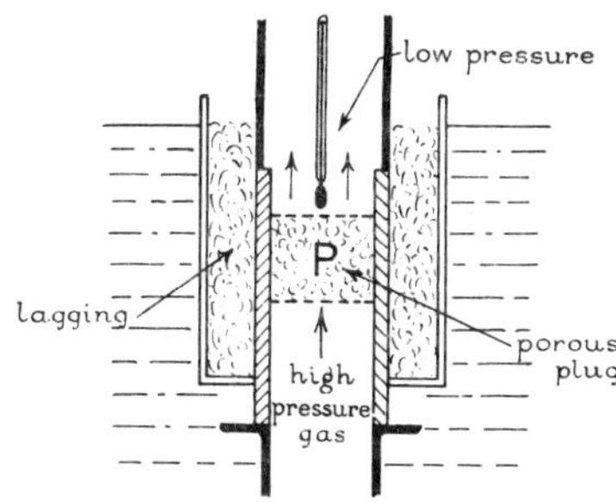

Fig. 6.15 Joule-Kelvin experiment on internal energy

adiabatic expansions of the gas, and prevented eddy currents in the gas which would produce undesirable energy changes. The plug was placed between two perforated brass plates in a vertical insulated boxwood (non-conducting) cylinder so that the gas expansion was adiabatic. The gas flowed through the plug into a region of low pressure, about one atmosphere, the high pressure on the other side of the plug being 4·5 atmospheres. The temperatures before and after passing through the plug were measured when conditions were steady, in which case all the heat change was expended on the gas alone and not on the container or thermometers. Joule and Kelvin, who worked with air, nitrogen, carbon dioxide, oxygen, and hydrogen, found *there was a slight cooling effect for all these gases except hydrogen*, which showed a slight heating effect. The temperature change was roughly proportional to the difference in pressure. The heating or cooling effect is known as the *Joule-Kelvin effect*. Experiments also show that the cooling effect diminishes as the temperature rises, and that above a certain temperature, known as the *inversion temperature*, a heating effect takes place. The inversion temperature varies with different gases, being about −80 °C for hydrogen and about −240 °C for helium.

The cooling on adiabatic expansion in the Joule-Kelvin effect should be carefully distinguished from that which occurs in adiabatic expansion of an ideal gas. In the latter case, *external* work is done. Thus even in the absence of intermolecular forces, which we assume to be the case for an ideal gas, energy is taken from the store of internal energy of the gas, whose temperature therefore falls. On the other hand, the Joule-Kelvin effect is due to *internal* work, that is, work done against intermolecular forces when the molecules become relatively farther apart. The Joule-Kelvin effect is thus zero for an ideal gas. It occurs only with real gases.

Linde method of liquefaction

In 1896 Linde liquefied air by using the Joule-Kelvin effect. A simple form of liquefier is illustrated diagrammatically in fig. 6.16(i). The air is first purified by passing it through tubes containing calcium chloride and lime, which remove traces of water-vapour and carbon dioxide.

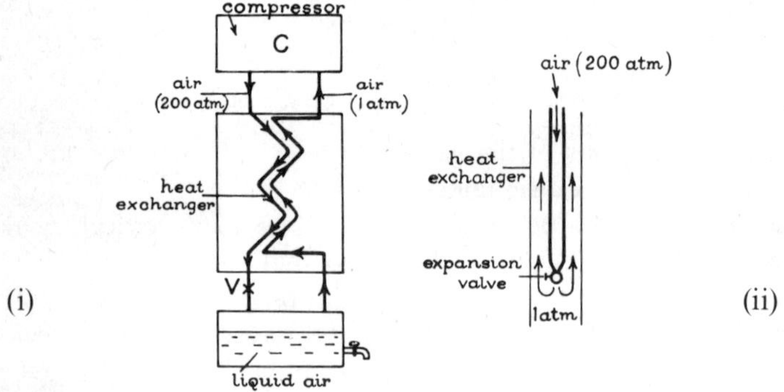

Fig. 6.16 Linde method of liquefaction—air

It is now compressed to about 200 atmospheres and the heat generated is removed by passing the air through a water-bath. It then expands through a valve V to 1 atmosphere. It is now cooled owing to the Joule-Kelvin effect. Some of this cooled air returns and circulates in a concentric pipe round the downflowing air, thus cooling it further before expansion. The arrangement, a so-called *heat exchanger*, is shown in fig. 6.16 (ii). The downcoming air thus becomes progressively cooled, and at some stage the critical temperature of air (−190 °C) is reached and passed. Liquid air is then obtained after expansion through V. It can be seen that the basic elements of a Linde liquefier are: a compressor, an expansion valve and a heat exchanger.

Linde improved the efficiency of the liquefier by cooling the air in two stages. The air is first cooled by allowing it to expand from 200 atm to 40 atm. Most of the air is then recirculated by a heat exchanger to give further cooling. About 20% of the cooled air expands through a second valve from 40 atm to 1 atm, when liquefaction is obtained.

Liquefaction of hydrogen

If a cooling effect is required for hydrogen when it undergoes a Joule-Kelvin effect, it must first be *pre-cooled* below about −80 °C or 193 K, its inversion temperature (see p. 145). Liquid nitrogen or liquid air is used for this purpose.

Figure 6.17 illustrates diagrammatically the basic principles of a simple small hydrogen liquefier. The gas is first compressed to 150 atmospheres. It is then passed through a charcoal cleaner which absorbs impurities, otherwise the system would be blocked when these solidify. Here it is also cooled by liquid nitrogen, which boils at −196 °C (77 K). It then passes another stage of pre-cooling at −213 °C (60 K), where the pressure may now be 130 atmospheres.

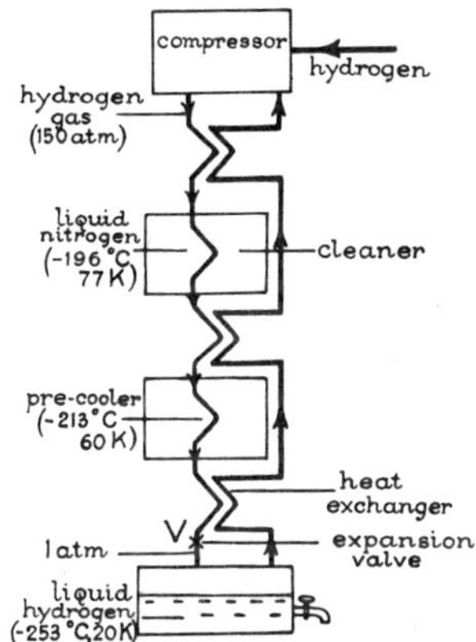

Fig. 6.17 Liquefaction of hydrogen

This temperature is below the inversion temperature of hydrogen. Thus on expansion through a valve V to a pressure of 1 atmosphere, cooling occurs and some of the gas becomes liquefied. A heat exchanger helps to cool the downflowing gas.

The basic principle is thus compression, pre-cooling below the inversion temperature, and expansion through a valve to low pressure, with the inclusion of a heat exchanger.

Claude's (external-work) method for liquefaction of air

When a gas does external work, e.g. in driving the piston of an engine, the gas becomes cooled if no heat is supplied to it, i.e. if it does work under adiabatic conditions. This is the basis of another method, used by Claude, for the liquefaction of air and other gases, and the *principle* of the method is illustrated in fig. 6.18.

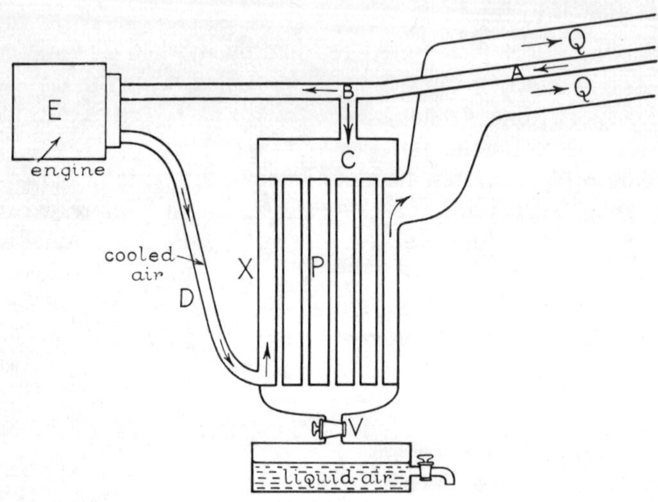

Fig. 6.18 Claude method of liquefaction

The gas to be liquefied, e.g. air, is compressed to about 40 atmospheres. The heat produced by the compression is then removed, and the air enters A and divides at B into two parts. About four-fifths of the air passes to an engine E, where it expands to one atmosphere and does external work in driving the engine. The air consequently becomes cooled, is returned by E through D to a chamber containing pipes P, and then passes back through Q to the compressor. The pipes P pass one-fifth of the air from A, and this is progressively cooled by the cold air passing round it in X. In this way the temperature becomes much lower than the critical temperature of air. The air entering the apparatus then liquefies, and is drawn off through the tap V.

The *efficiency* of the Claude process of liquefying air, i.e. the ratio of the volume of liquid air produced per hour to the volume of compressed air received per hour, is about the same as that obtained by Linde's method.

EXERCISE 6

Vapours

1. Distinguish between a *saturated* and an *unsaturated* vapour. Describe how the variation of the saturation vapour pressure of water-vapour with temperature may be investigated over a range from about 30 °C to 100 °C.

A closed vessel of fixed volume contains air and water. The pressures in the vessel at 20 °C and 75 °C are respectively 737·5 mm and 1144 mm of mercury, and some of the water remains liquid at 75 °C. If the saturation vapour pressure of water at 20 °C is 17·5 mm of mercury, find its value at 75 °C. (*L.*)

2. Distinguish between evaporation and boiling.

A beaker containing water is put under a bell-jar which is then evacuated. Describe and explain what is observed during the process. (*N.*)

3. State *Boyle's law* and *Dalton's law of partial pressures.*

The space above the mercury in a Boyle's law apparatus contains air together with alcohol vapour and a little liquid alcohol. Describe how the saturation vapour pressure of alcohol at room temperature may be determined with this apparatus.

A mixture of air and unsaturated alcohol vapour in the presence of liquid alcohol exerts a pressure of 128 mm of mercury at 20 °C. When the mixture is heated at constant volume to the boiling-point of alcohol at standard pressure (i.e. 78 °C), the vapour remaining saturated, the pressure becomes 860 mm of mercury. Find the saturation vapour pressure of alcohol at 20 °C. (*L.*)

4. What is meant by *saturation vapour pressure*? Describe an experiment to investigate the variation of the saturation pressure of water-vapour with temperature.

Sketch the isothermal curve relating pressure and volume (*a*) for a mass of dry air at room temperature, (*b*) for water vapour at 100 °C. (*L.*)

5. Use the simple kinetic theory of matter to answer the following questions: (*a*) How do gases conduct heat? (*b*) Why does a liquid tend to cool when it evaporates? (*c*) Why does the boiling-point of a liquid depend upon the external pressure?

Show that the pressure p, the density ρ, and the mean square molecular velocity $\overline{c^2}$ of an ideal gas are related by $p = \frac{1}{3}\rho\overline{c^2}$, stating any assumptions at the points where they become necessary in the proof. (*O. and C.*)

6. The saturation pressure of water-vapour is 12 mm of mercury at 14 °C and 24 mm of mercury at 25 °C. Describe and explain the experiment you would perform to verify these data.

Explain qualitatively in terms of the kinetic theory of matter (*a*) what is meant by saturation vapour pressure, (*b*) the relative magnitudes of the saturation vapour pressures quoted.

Sketch a graph showing how the saturation pressure of water-vapour varies between 0 °C and 110 °C. (*N.*)

7. Distinguish between a *saturated* and an *unsaturated* vapour. Describe an experiment to investigate the effect of pressure on the boiling-point of water and draw a sketch graph to show the general nature of the results to be expected.

A column of air was sealed into a horizontal uniform-bore capillary tube by a water index. When the atmospheric pressure was 762·5 torr (mm of Hg) and the temperature was 20 °C, the air-volumn was 15·6 cm long; with the tube immersed in a water bath at 50 °C it was 19·1 cm long, the atmospheric pressure remaining the same. If the s.v.p. of water at 20 °C is 17·5 torr, deduce its value at 50 °C. (*O. and C.*)

8. Describe in detail the method you would use to find (*a*) the melting-point of lead, and (*b*) the boiling-point of brine.

An alloy of copper and silver is made with different percentage composition. For each mix the melting-point is measured, with the following results:

M.P., °C	960	810	760	740	760	790	900	1080
Per cent of copper in alloy	0	20	30	40	50	60	80	100

Plot a curve showing the relation between the melting-point in degrees Centigrade and the percentage of copper present. Comment on the curve. (*L.*)

9. Compare the properties of saturated and unsaturated vapours. By means of diagrams show how the pressure of (*a*) a gas, and (*b*) a vapour, vary with change (i) of volume at constant temperature, and (ii) of temperature at constant volume.

The saturation vapour pressure of ether vapour at 0 °C is 185 mm of mercury and at 20 °C it is 440 mm. The bulb of a constant-volume gas thermometer contains dry air and sufficient ether for saturation. If the observed pressure in the bulb is 1000 mm at 20 °C, what will it be at 0 °C? (*L.*)

10. Explain concisely *four* of the following in terms of the simple kinetic theory of matter:

(*a*) energy must be supplied to a liquid to convert it to a vapour without change of temperature;

(*b*) when some water is introduced into an evacuated flask, some of the water at first evaporates, but subsequently, provided the temperature of the flask is kept constant, the volume of the water present remains unchanged;

(*c*) gases are generally poor conductors of heat compared with solids;

(*d*) when a gas, which is enclosed in a thermally insulated cylinder provided with a piston, is compressed by moving the piston, the temperature of the gas is raised;

(*e*) water can be heated by stirring. (*O. and C.*)

11. Define *dew-point* and explain what is meant by *relative humidity*. Describe how you would determine the dew-point of the atmosphere.

What is the relative humidity of an atmosphere whose temperature is 16·3 °C if its dew-point is 12·5 °C? The following table gives the saturation pressure p of water-vapour in mm of mercury at various temperatures t.

t °C	10	12	14	16	18	20
p mm	9·20	10·51	11·98	13·62	15·46	17·51

(*N.*)

Real gases. Critical temperature

12. State the postulates on which the simple kinetic theory of gases is based. What modifications are made to the postulates in dealing with real gases? How are these modifications represented in van der Waals' equation? (*N.*)

13. Describe the arguments which justify the van der Waals equation of state of a real gas:

$$\left(p + \frac{a}{V^2}\right)(V - b) = R\theta$$

If R in the above equation is the gas constant per unit mass, p is the pressure and θ the absolute temperature of the gas, what are the dimensions of the quantities V, a and b which appear in the equation?

A vessel of negligible heat capacity is thermally isolated from its surroundings and contains an ideal gas initially at high pressure. When a valve is opened to allow the gas to escape into the atmosphere the temperature of the gas which remains in the cylinder falls. Explain why this is so, and discuss the factors on which the final temperature of this gas depends. (*O. and C.*)

14. Define *pressure of a saturated vapour, critical temperature.* Give an account of the isothermal curves for carbon dioxide at temperatures above and below its critical temperature.

Some liquid ether is sealed in a thick-walled glass tube, leaving a space containing only the vapour. Describe what is observed as the temperature of the tube and its contents is raised above 197 °C, which is the critical temperature for ether. (Assume that the vessel is strong enough to withstand the internal pressure.) (*L.*)

15. Describe experiments in which the relation between the pressure and volume of a gas has been investigated at constant temperature over a wide range of pressure. Sketch the form of the isothermal curves obtained.

Explain briefly how far van der Waals' equation accounts for the form of these isothermals. (*L.*)

16. Describe, with a diagram, the essential features of an experiment to study the departure of a real gas from ideal gas behaviour.

Give freehand labelled sketches of the graphs you would expect to obtain on plotting (*a*) pressure P against volume V, (*b*) PV against P for such a gas at its critical temperature and at one temperature above and one below the critical temperature.

Explain van der Waals' attempt to produce an equation of state which would describe the behaviour of real gases.

Show that van der Waals' equation is consistent with the statement that all gases approach ideal gas behaviour at low pressures. (*O. and C.*)

17. Distinguish between an *isothermal* and an *adiabatic* compression of a gas. Explain the precautions necessary to ensure that an actual compression approximates to each of these conditions.

Sketch curves to show the relationship between pressure and volume, determined under isothermal conditions at a number of temperatures between about 10 °C and 40 °C for (*a*) carbon dioxide, (*b*) hydrogen. Explain the form of the curves.

Give a brief description of a method for liquefying hydrogen, emphasizing the physical principles involved rather than the technical details. (*C.*)

18. Explain what is meant by (*a*) a saturated vapour and (*b*) an unsaturated vapour. Draw a graph showing the form of the relation at various temperatures between the pressure and volume of a typical gas which liquefies, and indicate on your graph (i) the saturated vapour pressure at various temperatures and (ii) the critical point. Why is the compressibility of a liquid far less than that of a gas?

A vessel of volume V is maintained at a temperature T_1 and is connected by a tube of negligible volume to a second vessel maintained at a temperature T_2 which is higher than T_1. The system is evacuated and a liquid is then introduced into it. If the liquid has a saturated vapour pressure of P_1 at T_1, and P_2 at T_2, what would the pressure in the system be if the volume of the liquid present in it were (i) less than V, (ii) greater than V but less than the total volume of the system? Give reasons for your answers. (*O. and C.*)

19. Show that the density ρ of a perfect gas at pressure p and centigrade (Celsius) temperature t may be expressed in the form $\rho = \alpha p/(t+\beta)$, where α and β are constants. To what extent do the constants α and β depend on the nature of the gas?

Explain what is meant by an isothermal process, and sketch a number of isothermal lines for a perfect gas on a diagram in which ρ is plotted as ordinate and p as abscissa.

Describe briefly how the isothermal behaviour of a real gas may be studied experimentally at temperatures round the critical temperature and show how the isothermal lines on a graph of ρ against p differ from those you have drawn for a perfect gas. (*O. and C.*)

20. Derive an expression for the pressure (p) exerted by a mass of gas in terms of its molecular velocities, stating the assumptions made. What further assumption regarding the absolute temperature (T) of the gas is necessary to show that the expression is consistent with the equation $pv = kT$ where v is the volume of the mass of gas and k a constant? Which of the assumptions referred to did van der Waals modify to bring the equation $pv = kT$ more closely in agreement with the behaviour of real gases and what equation did he deduce? (*L.*)

21. State Boyle's law and describe how you would attempt to discover whether air shows any deviations from the law. Draw an approximate set of curves to show the way in which a gas deviates from Boyle's law in the region close to where the gas liquefies.

Two one-litre flasks are joined by a closed tap and the whole is held at a constant temperature of 50 °C. One flask is evacuated and the other contains air, water-vapour, and a small quantity of liquid water. The total pressure in the latter flask is 200 mmHg. The tap is then opened, and the system is allowed to reach equilibrium, when some liquid water remains. Assuming that air obeys Boyle's law, find the final pressure in the flasks. [Vapour pressure of water at 50 °C = 93 mmHg.] (*O. and C.*)

22. Explain the following in terms of the simple kinetic theory without mathematical treatment:

(*a*) A gas fills any container in which it is placed and exerts a pressure on the walls of the container.

(*b*) The pressure of a gas rises if its temperature is increased without the mass and volume being changed.

(*c*) The temperature of a gas rises if it is compressed in a vessel from which heat cannot escape.

(*d*) The pressure in an oxygen cylinder falls continuously as the gas is taken from it, while the pressure in a cylinder containing chlorine remains constant until very nearly all the chlorine has been used. The contents of the cylinder are kept at room temperature in both cases. [Critical temperature of chlorine = 146 °C.] (*C.*)

23. Give briefly the principles of one method each for liquefying (*a*) chlorine, (*b*) hydrogen.

Describe a suitable container for liquid air. Point out carefully the physical principles involved. (*C.*)

24. Give an account of three methods used to obtain temperatures below 0 °C. Describe a method for liquefying air. How must the procedure be modified in order to liquefy hydrogen? (*N.*)

25. A gas is compressed isothermally. Draw and discuss curves showing how its volume varies with its pressure, (*a*) if the temperature is above the critical temperature, (*b*) if the temperature is below the critical temperature. Describe and explain a method of liquefying air. (*N.*)

7

Thermal Conduction

Conduction of heat

If a teaspoon is placed in a cup of hot tea, the other end of the spoon soon becomes warmer. We say that some heat has passed, or been "conducted", along the metal from one end to the other.

The phenomenon of conduction of heat can be understood by considering the metal of a poker. Each of its molecules oscillates about a particular mean or "anchored" position (see p. 54). When one end of the poker is placed in a fire, thermal energy is passed to it and the molecules at this end become more energetic. The amplitudes of the oscillations then increase. The vibrating molecules now jostle or disturb the vibrating molecules in the next section more than before and give them additional energy. The increased amplitude of oscillation then affects molecules in the neighbouring section, and so on. Thus thermal energy is transferred from one section of the metal to the next, until the other end of the poker is reached. It should be carefully noted that the *average* position of the molecules remains unchanged while the transfer of energy takes place. The *conduction* of heat may hence be defined as *the transfer of heat from one part of a substance to another in which the average position of the intervening material remains constant*.

In the case of a metal, it may also be noted that free electrons appear to take part in conduction of thermal energy at normal temperatures (see p. 161).

Steady state in lagged bar

Consider a thick metal rod AB, with one end maintained at a constant temperature by steam, for example, and the other maintained at a lower constant temperature, say room temperature. The bar is "lagged", i.e. surrounded by layers of cotton-wool which prevent any heat escaping from the surface of the bar, and the bulbs of thermometers are placed in holes drilled in the bar at equal intervals to measure the temperatures of the different sections of the bar.

Before the steam circulates round the end A of the bar, the thermometers indicate that the temperatures of all the sections are the same.

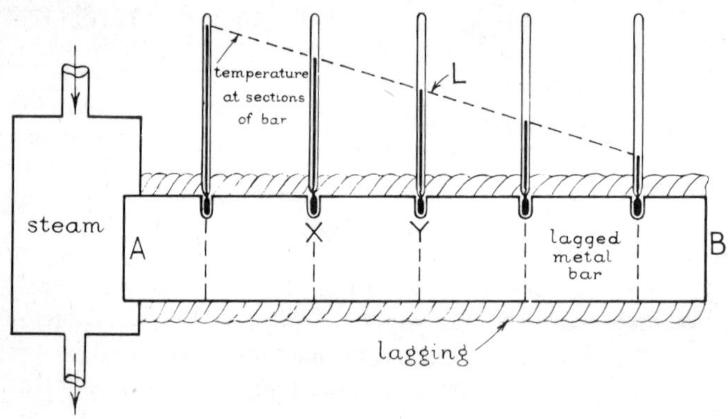

Fig. 7.1 Conduction in steady state—note the temperature gradient

When the steam passes round A, heat flows from this end through the bar and observation shows that the readings on the thermometers begin to alter. After some time, however, the temperature of each cross-section of the bar is observed to be constant, and a *steady state* is now said to exist along the bar AB. Experiment shows that, in the case of a lagged bar, the temperatures of the sections decrease *uniformly* with the distance from the hottest end. This is shown by the straight line L, which passes through the readings on each thermometer (fig. 7.1).

Fundamental formulae in conduction

In all experiments relating to conduction of heat, measurements are taken only when the steady state in the substance is reached, i.e. when the temperatures of its sections are constant, as in the case just considered. Since no heat is lost from the surface of the lagged bar in fig. 7.1, and the temperature of each section of the bar is constant, we can say that, in the steady state, the quantity of heat flowing per second through the section at X must be equal to that flowing through the section at Y and, in fact, to the quantity of heat flowing per second through any section of the bar.

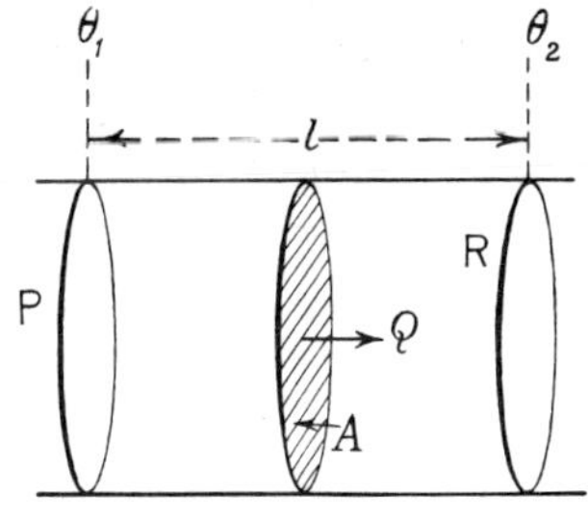

Fig. 7.2 Fundamental relation in conduction

Experiment shows that the quantity of heat Q flowing through a small part PR of a lagged bar of uniform cross-section in the steady state is proportional to each of three quantities (fig. 7.2). They are:

(*a*) *the time t*, i.e. the longer the time, the greater is the quantity of heat flowing through a given section;

(*b*) *the area A of the cross-section*, i.e. the larger the area the greater is the quantity of heat flowing through the section in a given time;

(*c*) *the temperature gradient g* established between the ends of PR. The temperature gradient g is defined as the ratio

$$\frac{\text{temperature difference between P and R}}{\text{length PR}}$$

Thus $g = (\theta_1 - \theta_2)/l$, where θ_1, θ_2 are the respective temperatures of P and R, and l is the length PR. If $\theta_1 = 75°\text{C}$, $\theta_2 = 68°\text{C}$ and $l = 10$ cm $= 0{\cdot}1$ m, then

$$g = \frac{75 - 68}{0{\cdot}1} = 70\ °\text{C m}^{-1} = 70\ \text{K m}^{-1}$$

Thermal conductivity

The quantity of heat Q flowing through any section is thus proportional to tAg, or $Q \propto tA(\theta_1 - \theta_2)/l$.

$$\therefore Q = ktA\frac{\theta_1 - \theta_2}{l} \quad . \quad . \quad . \quad . \quad . \quad . \quad (1)$$

where k is a constant of the material called its *thermal conductivity*.

From (1), the rate of flow of heat $Q/t = kA(\theta_1 - \theta_2)/l$. Thus the quantity of heat per second flowing through a section is given by

$$\frac{Q}{t} = kA \times \text{(temperature gradient)} \quad . \quad . \quad . \quad . \quad (2)$$

Although the relations (1) and (2) were introduced by consideration of steady heat flow, they apply to every case of heat flow by conduction, whether steady or unsteady heat flow. In terms of the calculus, relation (2) may be written generally as

$$\frac{dQ}{dt} = -kA\frac{d\theta}{dl} \quad . \quad . \quad . \quad . \quad . \quad . \quad (3)$$

where θ is the temperature at distance l along the conductor, so that $d\theta/dl$ represents the gradient to the particular $\theta - l$ curve at a section of the conductor and dQ/dt represents the rate of flow of heat through this section. The minus indicates that the temperature θ along a bar diminishes as l increases (see fig. 7.1), so that $d\theta/dl$ has a negative value and dQ/dt is then positive. Equation (3) can also be written as

$$\frac{1}{A}\cdot\frac{dQ}{dt} = -k\frac{d\theta}{dl} \quad . \quad . \quad . \quad . \quad . \quad . \quad (4)$$

or the heat flow per unit area per second = $k \times$ temperature gradient numerically.

In practice k may vary with direction in a particular material, i.e. the material is *anisotropic*. This is the case in crystals. Tables of constants quote different values of k for directions parallel and perpendicular to the principal axes of crystals. We have assumed in equations (1) to (4) that k has the same value for all directions so that we are dealing with an isotropic material; k also varies with temperature. In this book we discuss only isotropic materials and we assume k is constant in the temperature range concerned.

Definition and units of thermal conductivity

The thermal conductivity of a material may be defined as the constant k which appears in the relations $Q/t = kA \times$ (temperature gradient) or $dQ/dt = -kA d\theta/dl$. Alternatively, the thermal conductivity of a material may be defined as *the rate of flow of heat per unit area of cross-section per unit temperature gradient*; the flow of heat is here normal to the area and in the direction of the temperature gradient.

Since $$k = \frac{Q/t}{A \times \text{temperature gradient}}$$

the unit of k is $\dfrac{\text{joule/second}}{\text{metre}^2 \times {}^\circ\text{C/metre}}$, or $\dfrac{\text{joule/second}}{\text{metre} \times {}^\circ\text{C}}$

Using 1 K = 1 °C and 1 watt (W) = 1 joule per second, the SI unit of k is thus W m^{-1} K^{-1}.

Values of *k*

Experiment shows that, at normal temperatures, the thermal conductivity k of copper, a very good conductor, is about 380 W m^{-1} K^{-1}. Glass is a poor conductor; its value of k is about 1 W m^{-1} K^{-1}. Air is a very poor conductor; its value of k is about 0·03 W m^{-1} K^{-1}. The table below shows some values of the thermal conductivities of solids, liquids and gases.

SOLIDS (20 °C)

k (W m^{-1} K^{-1})			
silver	418	glass, window	1·0
copper	380	brick, building	0·6
iron	73	cardboard	0·2
tin	62	paper	0·06
germanium	60	wall board	0·06
steel, mild	47	felt	0·04
carbon	2	plywood	0·01

LIQUIDS (20 °C)

mercury	8
water	0·6
methyl alcohol	0·2
cylinder oil	0·15

GASES (20 °C)

air	0·03
oxygen	0·02
hydrogen	0·17
nitrogen	0·03
water-vapour	0·02

Generally, pure metals are good conductors of heat. Air, a poor conductor, is used in double glazing to prevent heat losses by conduction.

Examples

1. The total surface area of windows in a room is 2·5 metre^2 and the glass thickness is 3 mm. Calculate the heat lost per hour through the windows when the temperature inside the room is 20 °C and the temperature outside is 10 °C. Assume k = 1 W m^{-1} K^{-1} for glass.

$$\text{We have temperature gradient} = \frac{20 - 10}{3 \times 10^{-3}} \text{K m}^{-1}$$

$$\therefore Q/t = kA \times \text{temperature gradient}$$

$$= 1 \times 2{\cdot}5 \times \frac{20 - 10}{3 \times 10^{-3}} \text{J s}^{-1}$$

$$\therefore Q \text{ per hour} = 1 \times 2{\cdot}5 \times \frac{20 - 10}{3 \times 10^{-3}} \times 3\,600 \text{ J}$$

$$= 3 \times 10^7 \text{ J} = 3 \times 10^4 \text{ kJ}$$

2. Define *thermal conductivity* and describe *briefly* the essentials of a method of measuring this quantity for a good conductor.

The temperature inside a boiler is 105 °C. The wall of the boiler is 2 cm thick and is lagged with 4 cm thickness of a material whose thermal conductivity is 1/9 of that of the material of the boiler. When in the steady state the temperature of the outside surface of the lagging in contact with the air is 10 °C. What is the temperature of the common surface of the boiler and the lagging? (*L.*)

Let θ = temperature in °C of the common surface of the boiler and the lagging. The temperature gradient across the wall of the boiler is then $(105 - \theta)/0{\cdot}02$ K m^{-1} (2 cm = 0·02 m); the temperature gradient across the lagging is $(\theta - 10)/0{\cdot}04$ K m^{-1}. Thus if A is the area of the wall and also the area of the lagging material, and k and $k/9$ are the respective thermal conductivities, the quantities of heat passing per second through the wall and the lagging in the steady state are respectively

$$kA\frac{105 - \theta}{0{\cdot}02} \quad \text{and} \quad \frac{k}{9}A\frac{\theta - 10}{0{\cdot}04}$$

But in the steady state, the quantity of heat passing per second through the wall = the quantity of heat passing per second through the lagging.

$$\therefore kA\frac{105 - \theta}{0{\cdot}02} = \frac{k}{9}A\frac{\theta - 10}{0{\cdot}04}$$

$$\therefore \frac{105 - \theta}{2} = \frac{\theta - 10}{36}$$

$$\therefore 19\theta = 1900$$

$$\therefore \theta = 100 \text{ °C}$$

Effect of unburnt gas in boilers

When a boiler is heated, the metal plate at the bottom conducts the heat away from the hot gases, and their temperature falls below the ignition point. Thus a layer of unburnt gases forms on the underside of the boiler.

Although this layer is thin, a gas is such a poor conductor of heat that it has an appreciable effect on the flow of heat. As an illustration, suppose the gas is 1 mm thick and has a thermal conductivity of $1{\cdot}6 \times 10^{-2}$ W m^{-1} K^{-1} and that the steel plate at the bottom is 10 mm thick with a thermal conductivity of 40 W m^{-1} K^{-1}. Then, if θ is the temperature of the gas-metal boundary or underside of the boiler,

$$Q/t \text{ per metre}^2 = k_1 g_1(\text{gas}) = k_2 g_2(\text{metal})$$

where g_1, g_2 are the respective temperature gradients.

$$\therefore \frac{g_1}{g_2} = \frac{k_2}{k_1} = \frac{40}{1{\cdot}6 \times 10^{-2}} = 2500$$

Thus $g_1 = 2500g_2$, which shows that *a very steep drop of temperature occurs across the gas*, and only a small drop across the metal. Suppose θ is the temperature of the gas-steel boundary, and that 350 °C, 100 °C are the respective temperatures at the other sides of the gas and steel plate. Then, from above,

$$g_1 = \frac{350 - \theta}{0{\cdot}001} = 2500\frac{\theta - 100}{0{\cdot}01}$$

Solving, $\therefore \theta = 101$ °C

$$\therefore \text{flow of heat per m}^2 \text{ per second} = k_1 g_1 = 1{\cdot}6 \times 10^{-2} \times \frac{350 - 101}{0{\cdot}001}$$

$$= 4000 \text{ J}$$

With a copper plate in place of steel, which has a thermal conductivity k_3 of 360 W m^{-1} K^{-1}, the temperature gradient g_3 across it is given by

$$\frac{g_1}{g_3} = \frac{k_3}{k_1} = \frac{360}{1{\cdot}6 \times 10^{-2}} = 22\,500$$

A similar calculation to the above shows that the temperature of the gas-copper boundary is now about 100·1 °C.

$$\therefore \text{flow of heat per m}^2 \text{ per second} = k_1 g_1 = 1{\cdot}6 \times 10^{-2} \times \frac{350 - 100{\cdot}1}{0{\cdot}001}$$

$$= 4000 \text{ J (approx) as for steel.}$$

Thus the rate of flow of heat to the boiler is hardly affected by changing the boiler from steel to copper, although the latter is about 9 times the better conductor. As shown before, this is due to the very steep temperature fall across the layer of gas. The unburnt gas thus largely determines the rate of flow of heat to water in the boiler.

Growth of ice on ponds

Consider a layer of ice x cm thick formed on a pond, the temperature of the air just outside the ice being $-\theta$ °C. The temperature gradient across the ice is then θ/x. Suppose an extra thickness δx occurs in a time δt. Then, for an area A of the ice, a volume $A \, . \, \delta x$ of ice is obtained. The mass of ice is $\rho A \delta x$, where ρ is the density of the ice. The heat conducted away in this time $= lA\rho\delta x$, where l is the specific latent heat of fusion of ice, and hence

$$\frac{dQ}{dt} = lA\rho\frac{dx}{dt} = -kA\frac{-\theta - 0}{x} = kA\frac{\theta}{x}$$

$$\therefore \int_{x_0}^{x} x \, . \, dx = \frac{k\theta}{l\rho}\int_0^t dt$$

where x_0 is the thickness at $t = 0$ and x the thickness at time t.

$$\therefore \frac{x^2}{2} - \frac{x_0{}^2}{2} = \frac{k\theta}{l\rho}t$$

If $x_0 = 0$ when $t = 0$, that is, assuming no ice on the pond to start with,

$$\frac{x^2}{2} = \frac{k\theta}{l\rho}t$$

Hence in this case $x \propto \sqrt{t}$, or the thickness is proportional to the square root of the time. In practice, these results are very approximate.

Conduction of heat and electricity

An analogy may be made between the conduction of heat and the conduction of electricity. Thus (i) heat flow is due to a temperature gradient, current flow is due to a potential gradient; (ii) the heat flow is proportional to the temperature gradient, current flow is proportional to potential gradient or, for a given conductor, to the potential difference between its ends, provided the physical condition of the conductor remains unaltered (Ohm's law). Further, for heat flow,

$$\frac{Q}{t} = kA \times \frac{\text{temperature difference}}{l}$$

and for current flow, $I = Q/t = V/R = VA/\rho l$, since $R = \rho l/A$. Comparing the heat and current flow, it can be seen that

k is analogous to $1/\rho$

The quantity $1/\rho$ is defined as the *electrical conductivity* of the material.

Pure metals which are solid in the normal state, such as copper, aluminium and iron, are good conductors of both heat and electricity. This suggests that heat is conducted in metals at normal temperatures mainly by the transfer of energy by free electrons, which have thermal energy. Free electrons are also the carriers of charge in the conduction of electricity. In insulating solids, however, heat is conducted mainly by vibrations of molecules (p. 153).

Lagged and unlagged bars

If a bar is *lagged*, then all the heat which passes through one section per second will flow through other sections. In the steady state then, the rate of flow of heat, Q/t, is the same at each section. Now, at every section,

$$\text{temperature gradient} = \frac{Q/t}{kA}$$

It therefore follows that the gradient is *constant* along the whole length of the bar. Hence, for a lagged bar, the temperature falls linearly with distance from the hot end of the bar. This is shown by curve II in fig. 7.3 (i).

A different result is obtained for the *unlagged* or *bare bar*. Suppose a temperature gradient is set up in the direction AZ of the bar (fig. 7.3 (i)). In the steady state, let Q_1 be the quantity of heat entering

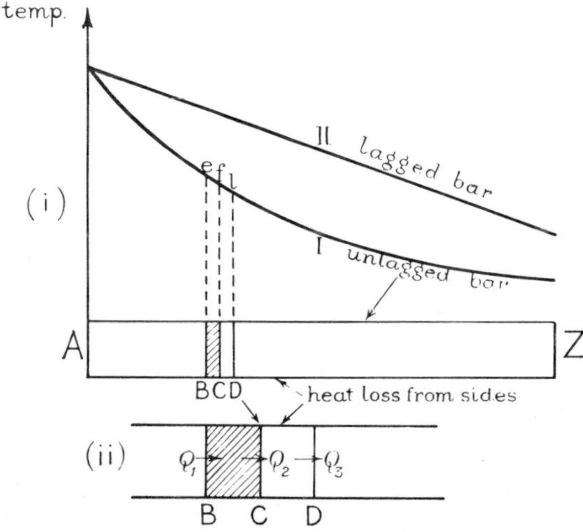

Fig. 7.3 Steady state in unlagged bar—compare lagged bar

the section B of the bar per second (fig. 7.3 (ii)). Since some heat is lost by radiation from the sides of BC, the quantity of heat per second Q_2 passing through the section C is less than Q_1. Similarly, the quantity of heat per second Q_3 passing through the section D is less than Q_2. On the average, then, the quantity of heat flowing per second through a section of the unlagged bar in the steady state becomes progressively smaller as we proceed from A to Z. Now the temperature gradient g across a small portion of the bar in the steady state is $\frac{Q/t}{kA}$ from equation (2) on page 156, where k and A are constants for a given bar. Hence, since Q/t decreases from A to Z, g decreases from A to Z. The temperature gradient thus diminishes along AZ, as shown by the curve I; the gradient of the temperature-distance curve between *fl*, for example, is less than the gradient between *ef*.

The heat capacity per unit volume of the bar = ρc, where ρ is the density of the material and c its specific heat capacity. Thus *before* the steady state is reached, the rate of temperature rise at any part of the bar is proportional to $k/\rho c$. The quantity "$k/\rho c$" is called the *diffusivity* of the material.

We can easily derive the variation of temperature θ with length l of the bar when it is lagged. In this case, in the steady state, $dQ/dt = a$, a constant.

$$\therefore -kA\frac{d\theta}{dl} = a$$

$$\therefore \int_{\theta_0}^{\theta} d\theta = -\frac{a}{kA}\int_0^l dl$$

where θ_0 is the temperature at $l = 0$. Integrating,

$$\therefore \theta - \theta_0 = -\frac{a}{kA}l$$

or

$$\theta = \theta_0 - \frac{a}{kA}l$$

This linear relation is shown in fig. 7.3 (i).

Radial heat flow

So far we have discussed linear or laminar heat flow; the heat flow through the windows of a warm room is practically linear flow. *Radial heat flow* occurs through metal pipes which carry hot water, as in a domestic heating system, or through lagging round circular tanks.

Consider a section of a cylindrical metal pipe carrying hot water at a constant temperature θ_1 (fig. 7.4). Suppose the external surface

C has a constant temperature θ_0. The heat flow from the interior metal surface H to C passes radially through all the cylindrical sections such as S. If S has a radius r, the area A of S is $2\pi rl$, where l is the length of the pipe. *Thus A varies with r.* It can now be seen that the heat flow per unit area through the different metal sections decreases as r increases. With linear flow for a lagged bar, however, the heat flow per unit area through the different sections is constant.

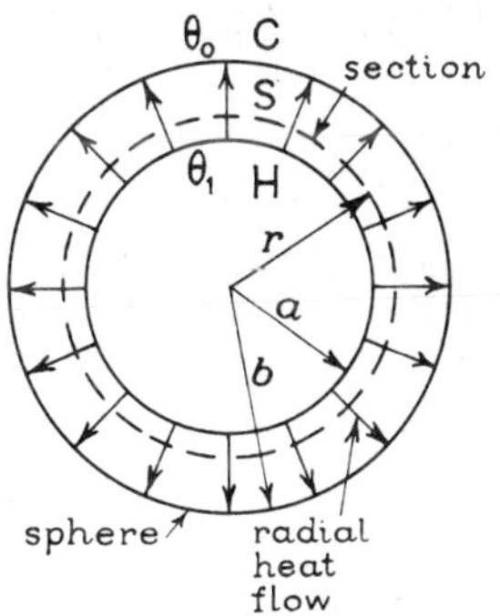

Fig. 7.4 Radial heat flow

Suppose a watts is the rate of flow of heat through the pipe in the steady state. Then

$$Q/t = a = k \,.\, A \times \text{(temperature gradient)}$$

$$\therefore a = k \,.\, 2\pi rl \,\text{(temperature gradient)} \quad . \quad . \quad . \quad (1)$$

Hence if the rate of supply of heat a is 100 W, $l = 1$ metre, $k = 400$ W m^{-1} k^{-1} and the radius r of the external surface C is 2·5 cm or $2{\cdot}5 \times 10^{-2}$ m, then, from (1),

$$\text{temperature gradient at C is } \frac{a}{k \,.\, 2\pi rl} = \frac{100}{400 \,.\, 2\pi \times 2{\cdot}5 \times 10^{-2} \times 1}$$

$$= 1{\cdot}6 \text{ K m}^{-1}$$

If a small heater is placed at the centre of a *spherical conductor*, radial heat flow this time occurs through spherical sections whose area is given by $4\pi r^2$, where r is the radius of the particular section. As an example, suppose the heater supplies 40 W, the thermal conductivity of the metal sphere is 100 W m^{-1} K^{-1}, and the temperature gradient g at a section of radius 5 cm is required when conditions are steady. Then, since $r = 5 \times 10^{-2}$ m, from $Q/t = kAg$ we have

$$40 = 100 \times 4\pi \times (5 \times 10^{-2})^2 \times g$$

$$\therefore g = \frac{40}{100 \times 4\pi \times 25 \times 10^{-4}} = 13 \text{ K m}^{-1}$$

Relation for k

We can now find the relation between k and the actual temperatures θ_1, θ_0 of the internal and external surfaces of a cylindrical pipe.

Using the calculus, we have from previous work on p. 163

$$a = -k \,.\, 2\pi rl \,.\, \frac{d\theta}{dr}$$

where $d\theta/dr$ represents the temperature gradient at the section S in fig. 7.4.

$$\therefore a \int_{r_1}^{r_2} \frac{dr}{r} = -2\pi kl \int_{\theta_0}^{\theta_1} d\theta$$

where r_1, r_2 are the respective radii of the interior and exterior surfaces.

$$\therefore a \log_e (r_2/r_1) = 2\pi kl\,(\theta_0 - \theta_1)$$

$$\therefore k = \frac{a}{2\pi\, l\,(\theta_0 - \theta_1)} \log_e \frac{r_2}{r_1} \quad . \quad . \quad . \quad . \quad . \quad (1)$$

When r_2 is slightly greater than r_1, as in a thin pipe or tube, $\log_e (r_2/r_1) = (r_2 - r_1)/r_1$ approximately. See also p. 169.

Suppose now that the steady supply of heat is at the centre of a *spherical conductor* of internal and external radii r_1, r_2 respectively. Then, as explained,

$$a = -k \,.\, 4\pi r^2 \,.\, \frac{d\theta}{dr}$$

$$\therefore a \int_{r_1}^{r_2} \frac{dr}{r^2} = -4\pi k \int_{\theta_0}^{\theta_1} d\theta$$

where θ_0, θ_1 are the respective internal and external temperatures of the sphere.

$$\therefore a\left(\frac{1}{r_1} - \frac{1}{r_2}\right) = 4\pi k(\theta_0 - \theta_1) \quad . \quad . \quad . \quad . \quad . \quad (2)$$

Radial heat flow in wire

As a further illustration of radial flow of heat, consider an electric current I flowing in a wire of electrical resistivity ρ, thermal conductivity k, radius a and length l (fig. 7.5). To find the temperature difference α between the axis of the wire and its surface, take a cylinder P of wire of radius r, as shown.

Then, since the *current density* in the wire is $I/\pi a^2$, the current through P $= I \times \pi r^2/\pi a^2 = Ir^2/a^2$.

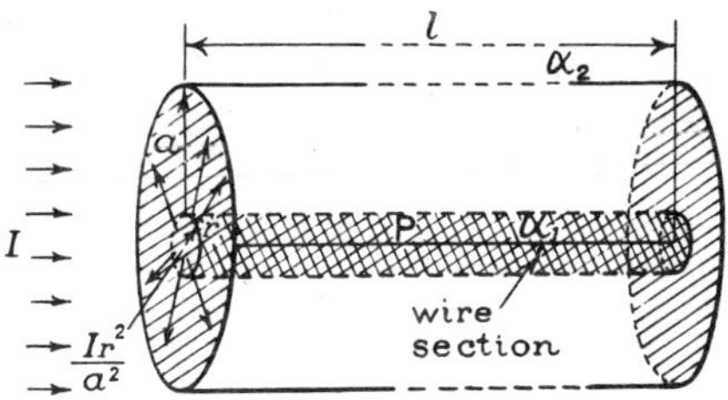

Fig. 7.5 Electrical heating of wire—radial flow

The resistance of P is $\rho l/\pi r^2$. The area through which heat flows is $2\pi rl$. Since heat per second $= -k \times$ area $\times$ temperature gradient.

$$\therefore \left(\frac{Ir^2}{a^2}\right)^2 \times \frac{\rho l}{\pi r^2} = -k \,.\, 2\pi rl \,.\, \frac{d\theta}{dr}$$

$$\therefore \frac{I^2 \rho l}{\pi a^4} \int_0^a r \,.\, dr = -2\pi kl \int_{\alpha_1}^{\alpha_2} d\theta$$

where α_1, α_2 are the respective temperatures at the axis and surface.

$$\therefore \frac{I^2 \rho}{4\pi^2 ka^2} = \alpha_1 - \alpha_2 = \alpha$$

Searle's method for k

Searle designed an apparatus for measuring the thermal conductivity of a good conductor such as a metal. A thick cylindrical bar S of the specimen, heated electrically at one end A by a coil, is well lagged by felt and enclosed (fig. 7.6). Holes are drilled into the bar at places E, H about 10 cm apart, and are filled with mercury so that the bulbs of thermometers placed in the holes are in good thermal contact with the sections of the bar at E and H. A copper coil C, soldered round the bar, has a steady flow of water passing in at D and out at B, and the inlet and outlet temperatures are measured by thermometers.

When the electrical heating coil is switched on, the temperatures of the thermometers begin to rise. After a time, the readings on the thermometers become constant, indicating that a *steady state* has been reached, and the readings are then taken. The steady flow of water through the coil C is measured by means of a beaker and a

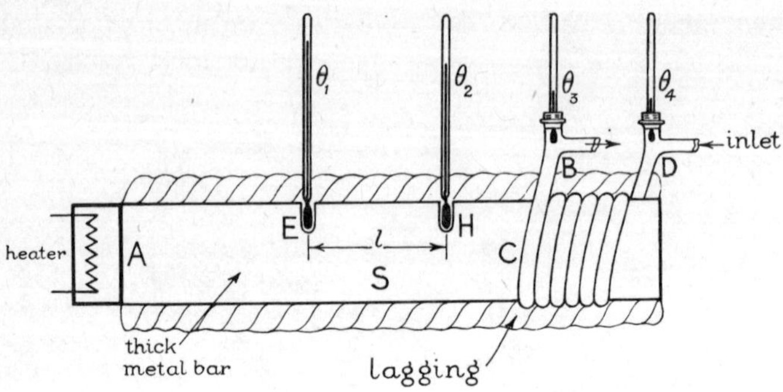

Fig. 7.6 Searle method for good conductor—lagging and uniform cross-section are essential

stopclock. If m is the mass of water collected in a time t, the quantity of heat per second (Q/t) flowing through the bar is $mc_W(\theta_3 - \theta_4)/t$, where θ_3, θ_4 are the respective temperatures at B and D. But $Q/t = kA \times$ temperature gradient $= kA(\theta_1 - \theta_2)/l$, where l is the length EH, A is the cross-sectional area of the bar, and θ_1, θ_2 are the respective temperatures at E and H.

$$\therefore kA\frac{\theta_1 - \theta_2}{l} = \frac{mc_W(\theta_3 - \theta_4)}{t}$$

Hence k can be calculated as the other quantities in the equation are known. The area of cross-section $A = \pi r^2$, where r is the radius of the cylindrical bar, and r can be measured by means of calipers.

In Searle's apparatus, the following points may be noted: (1) The quantity of heat per second through the bar is proportional to its area of cross-section. A thick bar is sufficient to provide a measurable quantity of heat in a reasonable time since a metal is a good conductor. (2) A bar about 20 cm long or more can provide a reasonable temperature gradient for a metal when one end is heated electrically or by steam. (3) Lagging is essential since the bar is long, otherwise heat would escape through the sides of the bar by radiation and the heat flow would not be linear.

Methods for bad conductors

We have now to consider various methods which have been devised to find the thermal conductivity k of bad conductors, such as cardboard, ebonite, glass, and rubber, for example. In order to obtain an appreciable quantity of heat flowing per second through the specimen in a steady state, (*a*) a high temperature gradient must be set up across it, and (*b*) a large area of cross-section must be used. This means, in practice, that a thin specimen of large cross-sectional area must be used in the experiment. Further, the heat lost per second in the steady state from the sides of the specimen must be very small, in which case the fundamental equation $Q/t = kA \times$ (temperature gradient) can be used for the flow of heat through the specimen.

Lees' disc method for bad conductor

Lees devised a method of measuring the thermal conductivity of a bad conductor in the form of a disc; the method is thus applicable to such materials as cardboard or ebonite. The disc X, which is thin, is sandwiched between two circular nickel-plated brass plates M, N, which are much thicker than X, and the upper plate M forms the bottom of a hollow cylinder P through which steam can be passed (fig. 7.7). The temperatures of M and N can be obtained from thermometers T_1, T_2 respectively, which are placed inside holes drilled in

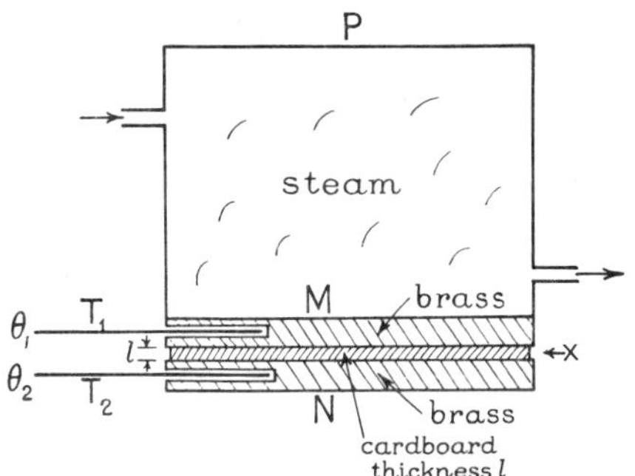

Fig. 7.7 Lees' disc method for bar conductor—note the high temperature gradient and large cross-sectional area

M, N. In the experiment, steam is passed into P, and the readings on T_1, T_2 are taken when they are constant, indicating that a steady state is obtained. Neglecting the amount of heat lost per second from the sides of X, which is very small because the surface area of the sides is very small, we have $Q/t = kA(\theta_1 - \theta_2)/l$, where Q/t is the quantity of heat per second flowing through X in the direction MN, A is the area of cross-section of X, k is its thermal conductivity, θ_1, θ_2 are the respective temperatures on T_1, T_2, and l is the small *thickness* of the disc.

A separate experiment is performed to find the quantity of heat flowing per second (Q/t) through the disc in the steady state, which is equal to the heat lost per second by radiation from the exposed surface of N. To find thc hcat lost by radiation, P and M are removed, leaving X above N; N is gently warmed by a burner until its temperature exceeds the value θ_2, and then N is allowed to cool. Readings of the temperature are taken at equal intervals of time until it has a value some degrees lower than θ_2, and the cooling curve is plotted. The rate of cooling in °C per second at the temperature θ_2 is then found by drawing the appropriate tangent to the curve; we shall suppose the value is a °C per second. As the conditions under which N has cooled are the same as in the first part of the experiment, the heat lost per second by N at θ_2 °C by cooling = the heat flowing per second Q/t through X in fig. 7.7. But the heat per second lost by N at θ_2 °C is mca, where m is the mass of N and c its known specific heat capacity.

$$\therefore \frac{Q}{t} = kA\frac{\theta_1 - \theta_2}{l} = mca$$

Knowing $m, c, a, A, \theta_1, \theta_2$, and measuring l with a micrometer screw-gauge, the thermal conductivity k can be calculated from this equation.

Method for bad conductor in form of tube

The thermal conductivity of a solid, such as glass, in the form of a tube, can be measured by the apparatus shown in fig. 7.8 (i). The

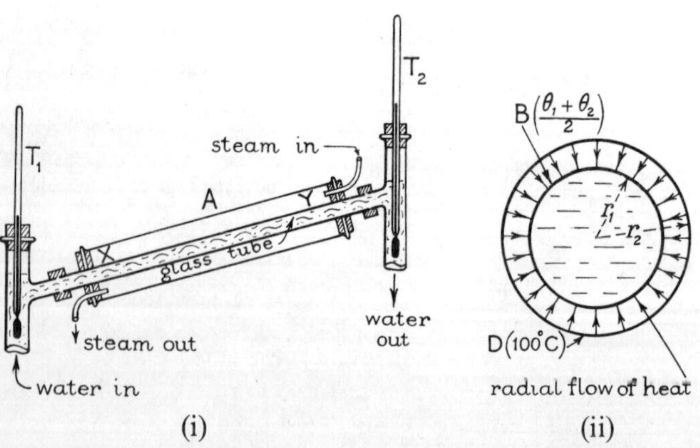

Fig. 7.8 Bad conductor in form of cylinder—radial heat flow

tube XY is surrounded by a jacket A, and water flows through the tube from X to Y at a steady rate. When the steam circulates round XY heat passes through the tube to the flowing water inside it, which therefore becomes warmed, and after a time the outlet temperature of the water, indicated by the reading on the thermometer T_2, becomes constant. A steady state is now obtained, and the temperature t_2 on T_2 is noted.

Suppose the length of XY is L, and r_1, r_2 are the internal and external radii of the tube respectively. The average radius r of the tube is then $\frac{1}{2}(r_1 + r_2)$, and the surface area A of the tube $= 2\pi rL = \pi(r_1 + r_2)L$. The temperature of the outside surface D of the tube is the temperature of the steam, 100 °C say, and the temperature of the interior surface B is the average temperature $\frac{1}{2}(\theta_1 + \theta_2)$ of the flowing water, where θ_1 is the inlet temperature of the water (fig. 7.8 (ii)). Consequently the temperature gradient across the thin tube = temperature difference / thickness $= [100 - \frac{1}{2}(\theta_1 + \theta_2)] / (r_2 - r_1)$. Hence the quantity of heat per second flowing from D to B in the steady state is given approximately (see p. 164) by

$$Q/t = kA\frac{100 - \frac{1}{2}(\theta_1 + \theta_2)}{r_2 - r_1} = k\pi(r_1 + r_2)L\frac{100 - \frac{1}{2}(\theta_1 + \theta_2)}{r_2 - r_1}$$

where k is the thermal conductivity of the material of the tube. This quantity of heat per second is $mc_W(\theta_2 - \theta_1)$, where m is the mass of water per second flowing through XY and c_W its specific heat capacity, since the water is warmed from a temperature θ_1 to θ_2.

$$\therefore k\pi(r_1 + r_2)L\frac{100 - \frac{1}{2}(\theta_1 + \theta_2)}{r_2 - r_1} = mc_W(\theta_2 - \theta_1)$$

Thus k can be calculated, as the other quantities in the equation are known.

A piece of thick twisted wire is often placed along the tube XY to mix the water flowing through it. It should be noted that the heat flows through the tube along the *radii* of the cross-section, as represented by the arrows in fig. 7.8 (ii), and this flow of heat is therefore called a *radial flow* (see p. 162). In distinction, the heat flows along the bar in fig. 7.6 in parallel directions, and this is termed a *uniform* or *laminar flow* of heat.

Thermal conductivity of rubber tube

The thermal conductivity of a rubber tube can be found by first passing steam through it for a short time. The tube is then picked up, with the steam still issuing from it, and a fixed length of it, XMY, is immersed inside a calorimeter containing water, which has been cooled several degrees below room temperature by the addition of ice (fig. 7.9(i)). The temperature of the water is then noted at equal intervals of time as it rises from its initial temperature, corresponding to P (fig. 7.9(ii)), to a temperature N above the room temperature θ_1; from this graph the rise in temperature per second, a °C per second say, at the temperature θ_1 is obtained by drawing the corresponding tangent l to the curve.

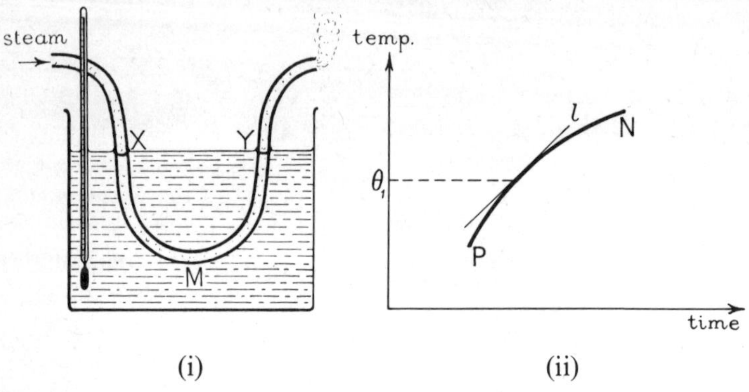

Fig. 7.9 Conductivity of rubber tubing

When the water and calorimeter are at the same temperature as the room (θ_1) no heat is lost by them to the surroundings. Thus the quantity of heat flowing per second (Q/t) through the rubber tube when immersed in the water is equal to the heat per second gained by the water and calorimeter at the temperature θ_1. Hence $Q/t = C.a$ where C is the total heat capacity of the water and the calorimeter.

$$\therefore \frac{Q}{t} = k\pi (r_1 + r_2)\, l \frac{\theta - \theta_1}{r_2 - r_1}$$

since the area A through which the heat is conducted is $\pi(r_1 + r_2)l$, and the temperature gradient is $(\theta - \theta_1)/(r_2 - r_1)$, where l is the length of XMY, r_1, r_2 are the external and internal radii of the tube, and θ is the temperature of the steam (compare the case of the follow tube, p. 168). Thus

$$Ca = k\pi(r_1 + r_2)l \frac{\theta - \theta_1}{r_2 - r_1}$$

The thermal conductivity k of the rubber can hence be calculated, as the other quantities in the equation are known.

Thermal conductivity of liquids

Apart from mercury, liquids are usually bad conductors of heat. In determining the thermal conductivity k of a liquid, the main precaution is to ensure that no convection can take place, as this would spoil the experiment, and for this purpose the temperature of the *top* of the liquid is made greater than the bottom.

Lees' disc method for bad solids (p. 167) can be adapted for measuring the thermal conductivity of a liquid. Four flat copper discs of about the same diameter, A, B, C, D, are set up horizontally with a heating coil between A and B, a glass slab G between B and C, and the liquid L in an ebonite ring E between C and D (fig. 7.10). When conditions are steady, the temperatures of B, C, D are measured by thermocouples, and since the thermal conductivity of the glass G had been previously determined, the quantity of heat flowing per

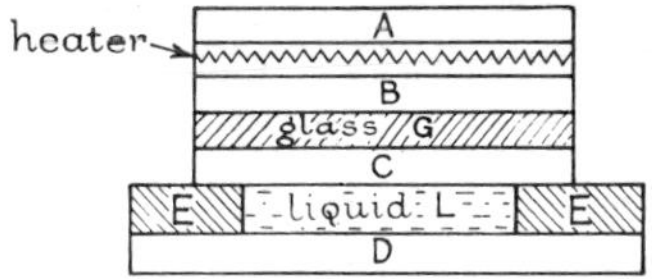

Fig. 7.10 Conductivity of liquid (C at higher temperature than D eliminates convection)

second down through C is known. By subtracting the heat flowing per second through E (determined by a subsidiary experiment with air in place of the liquid), the heat flowing per second through L is determined; and since the temperature difference between C and D, and the area of L are known, the thermal conductivity of the liquid can be calculated.

Thermal conductivities of gases

Gases are very poor conductors of heat, e.g. the thermal conductivity of air is about 0·03 W m^{-1} K^{-1}. On this account the film of gases on the under side of a boiler reduces considerably the temperature immediately underneath the boiler, and the scale formed must be cleaned from the boiler periodically (p. 158).

The task of determining the thermal conductivity of a gas is more formidable than for a liquid, as radiation as well as convection effects are liable to occur in the case of the gas. Figure 7.11 illustrates the

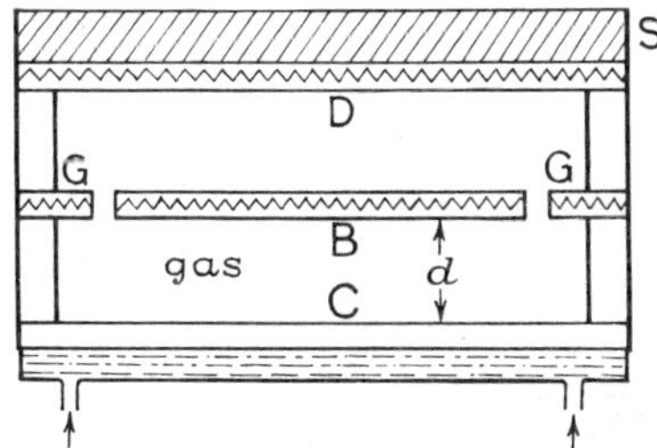

Fig. 7.11 Conductivity of gas (convection and radiation eliminated)

principle of a method for determining the thermal conductivity k of a gas. Three plane polished plates, D, B, C, are made of copper and silverplated, and C is cooled by a stream of water so that its temperature is constant. B is heated electrically by a coil inside it, and is surrounded by a ring G similarly heated; D is also heated electrically and has an insulator S above it. By adjusting the currents in D, G, B, the temperatures of these metals can be made the same, and after a steady state is reached the temperatures t_1, t_2 of B, C are recorded by means of thermocouples.

Since B and D are at the same temperature there is no convection between them. Further, as B is at a higher temperature than C, there is no convection between B and C. There is no radiation between B and D since they are at the same temperature, and the radiation between B and C is small since the plates are silverplated. No conduction takes place between D, B and G, B since they are at the same temperature, and the heat flows downward from B to C, the guard ring G ensuring that the flow of heat is perpendicular to B and C. The rate of heat supplied to B, and the temperatures of B and C are known, and since Q per second $= kA(\theta_1 - \theta_2)/l$, where l is the distance between B and C, and A is the area of B, the thermal conductivity k of the gas can be calculated.

Harder Examples on Conduction

1. Compare the laws governing the conduction of heat and of electricity, pointing out the corresponding quantities in each case. Outline an experiment to prove the heat law corresponding to Ohm's law in electricity.

A copper rod is lagged, so that no heat is lost through its sides, and is uniformly heated by the passage of an electric current. Show, by considering a small section δx, that the temperature T varies with distance x along the rod in such a way that $k(d^2T/dx^2) = -H$, where k is the thermal conductivity and H the rate of heat generation per unit volume. If the rod is 10 cm long, 1 cm in diameter and dissipates a total energy of 100 W, calculate the excess temperature at its centre when both ends are held at a constant temperature. (Thermal conductivity of copper = 400 W m^{-1} K^{-1}.) (*C.S.*)

(1) Suppose BC is a small section of the lagged copper rod, where B is a distance x

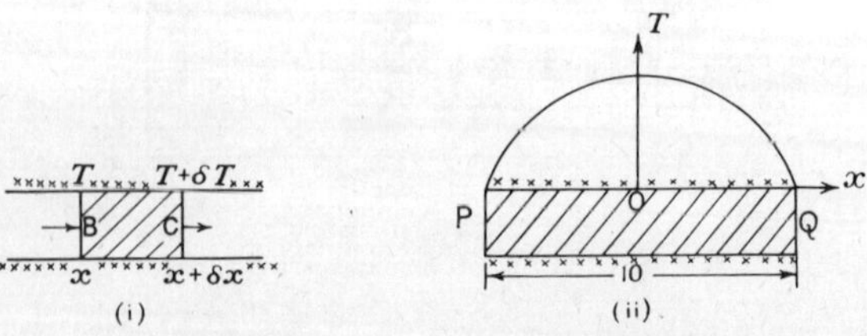

Fig. 7.12 Example

along the rod at a temperature T and C is a distance $x + \delta x$ at a temperature $T + \delta T$ (fig. 7.12(i)). Then, if A is the area of cross-section,

$$\text{rate of heat through B} = -kA\frac{dT}{dx} \qquad \text{(p. 156)}$$

and

$$\text{rate of heat through C} = -kA\frac{dT}{dx} + \frac{d}{dx}\left(-kA\frac{dT}{dx}\right)\delta x$$

$$= -kA\frac{dT}{dx} - kA\frac{d^2T}{dx^2}\delta x$$

Since the section BC is lagged, by subtraction,

$$\text{rate of heat generated in BC} = -kA\frac{d^2T}{dx^2}\delta x$$

But $A \,.\, \delta x$ = volume of element BC.

$$\therefore -k\frac{d^2T}{dx^2} = H \quad \text{or} \quad k\frac{d^2T}{dx^2} = -H \qquad \text{(i)}$$

(2) Since the rod PQ is 10 cm long and its radius is 0·5 cm (fig. 7.12(ii)),

$\therefore$ energy dissipated per unit volume per second

$$= \frac{100}{10 \times \pi \times 0{\cdot}5^2} = \frac{40}{\pi}\,\text{W cm}^{-3} = \frac{40}{\pi} \times 10^6\,\text{W m}^{-3}$$

From (i)

$$\therefore 400\frac{d^2T}{dx^2} = -\frac{40}{\pi} \times 10^6$$

$$\therefore \frac{d^2T}{dx^2} = -\frac{10^5}{\pi}$$

Integrating,

$$\therefore \frac{dT}{dx} = -\frac{10^5}{\pi}x + c_1 \qquad \text{(ii)}$$

With the origin of x at P, then $dT/dx = 0$ when $x = 0{\cdot}05$. Hence, from (ii), $c_1 = 5000/\pi$.

$$\therefore \frac{dT}{dx} = \frac{10^5}{\pi}(0{\cdot}05 - x)$$

$$\therefore T = \frac{10^5}{\pi}(0{\cdot}05x - \tfrac{1}{2}x^2) + c_2$$

Since $T = 0$ when $x = 0$, then $\quad c_2 = 0$

$$\therefore T = \frac{10^5}{\pi}(0{\cdot}05x - \tfrac{1}{2}x^2)$$

At the centre, $x = 0{\cdot}05$ $\quad \therefore T$ = excess temperature

$$= \frac{10^5}{\pi}(0{\cdot}0025 - \tfrac{1}{2} \times 0{\cdot}0025)$$

$$= 40\,^\circ\text{C (approx)}$$

2. Define the thermal conductivity k of a solid and describe an accurate method of measuring k for an extremely poor conductor such as expanded polythene.

A thin-walled copper sphere of radius 5 cm and mass 100 g containing 400 g water is cooled to -176 °C by immersing it in liquid air. It is then placed inside a fitting hollow sphere of expanded polythene of outer radius 10 cm in a room at 20 °C. What is the value of k if the ice just melts after 24 h? Assume the specific heat capacities of ice and copper remain constant at 2·1 and 0·4 kJ kg^{-1} K^{-1} respectively and the specific latent heat of ice is 336 kJ kg^{-1} K^{-1}. (*C.S.*)

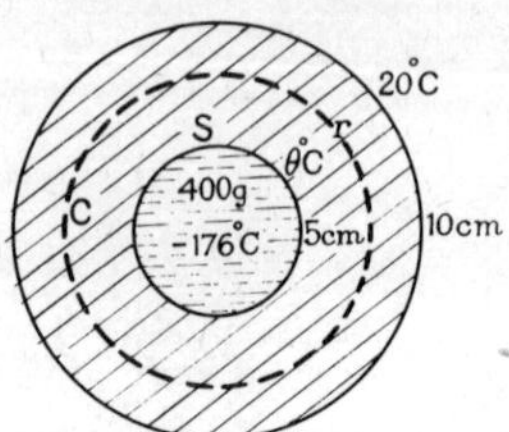

Fig. 7.13 Example

Consider a section C of the polythene of radius r, and suppose the surface S of the copper sphere rises from a temperature θ to $(\theta + \delta\theta)$ as heat is conducted through the polythene (fig. 7.13). Then, for conduction through the surface S,

$$\frac{dQ}{dt} = a = k \,.\, 4\pi r^2 \,.\, \frac{d\theta}{dr}$$

where a is the quantity of heat per second flowing while the temperature of S rises from θ to $(\theta + \delta\theta)$.

$$\therefore \int_{20}^{\theta} d\theta = \frac{a}{4\pi k}\int_{0.1}^{0.05} \frac{dr}{r^2}$$

$$\therefore \theta - 20 = \frac{a}{4\pi k}\left(\frac{1}{0{\cdot}1} - \frac{1}{0{\cdot}05}\right)$$

$$\therefore a = 0{\cdot}4\pi k(20 - \theta) \quad \text{(i)}$$

But

$$a = \frac{dQ}{dt} = (100 \times 0{\cdot}4 + 400 \times 2{\cdot}1)\frac{d\theta}{dt} \quad \text{(ii)}$$

$$\therefore 880\frac{d\theta}{dt} = 0{\cdot}4\pi k(20 - \theta)$$

Hence the time t to reach 0 °C is given by

$$t = \frac{2200}{\pi k}\int_{-176}^{0} \frac{d\theta}{20 - \theta} = \frac{2200}{\pi k}\log_e\left(\frac{196}{20}\right) \quad \text{(iii)}$$

After the surface of S reaches 0 °C, the rate of flow of heat a_1 along the polythene is constant. To find a_1, we have, from (i),

$$a_1 = 0{\cdot}4\,\pi k \,.\, 20 = 8\pi k$$

$$\therefore \text{time } t_1 \text{ for all ice just to melt} = \frac{ml}{8\pi k} = \frac{0{\cdot}4 \times 336\,000}{8\pi k} \quad \text{(iv)}$$

$$\therefore \text{total time} = t + t_1 = \frac{2200}{\pi k}\log_e\frac{196}{20} + \frac{16\,800}{\pi k}$$

$$= 24 \times 3600$$

$$\therefore k = \frac{1}{24 \times 3600}\left[\frac{2200}{\pi}\log_e\left(\frac{196}{20}\right) + \frac{16\,800}{\pi}\right]$$

$$= 8{\cdot}0 \times 10^{-2}\,\text{W m}^{-1}\,\text{K}^{-1}$$

EXERCISE 7

1. The opposite sides of a solid cube of side 5 cm are maintained at temperatures of (i) 100 °C and 15 °C, (ii) 20 °C and -5 °C. Calculate the temperature gradient in each case in K m^{-1} (°C m^{-1}).

2. Write down the expression for the quantity of heat per second flowing through a substance in the steady state. Two opposite faces of a solid copper cube of side 20 cm are maintained at temperatures of 85 °C and 5 °C respectively. Calculate the heat flowing through the cube in the steady state in 5 min. (Assume $k = 400$ W m^{-1} K^{-1} for copper.)

3. Show that W m^{-1} K^{-1} is the unit of *thermal conductivity*. The thickness of a glass window is 2 mm, and it has an area of 0·8 m^2. If one side of the glass has a temperature of 16 °C and the other has a temperature of -4 °C, calculate the quantity of heat flowing through the glass in 10 seconds. (Assume k for glass = 1·0 W m^{-1} K^{-1}.)

4. The ends of a long cylindrical bar are maintained at a constant difference of temperature. Draw a sketch of the temperature variation of the different sections of the bar in the steady state if it is (i) unlagged, (ii) lagged. Explain briefly the reason for the temperature variation in both cases.

5. How is the quantity of heat flowing per second through the bar in the steady state measured in Searle's method of determining thermal conductivity? What is the advantage of lagging the bar?

6. What difficulties arise in experimental methods of measuring the thermal conductivity of (i) liquids, (ii) gases? State briefly how the difficulties may be overcome.

7. Define *thermal conductivity*. Describe a method of measuring this quantity for a metal.

Assuming that the thermal insulation provided by a woollen glove is equivalent to a layer of quiescent air 3 mm thick, determine the heat loss per minute from a man's hand, surface area 200 cm^2 on a winter's day when the atmospheric air temperature is -3 °C. The skin temperature is to be taken as 34 °C and the thermal conductivity of air as 24×10^{-3} W m^{-1} K^{-1}. (*L.*)

8. Estimate in kWh the energy lost by conduction through the windows of a typical house on a cold December day in Manchester. Comment on your answer. Take the thermal conductivity of glass to be 1 W m^{-1} K^{-1}. (*N.*)

9. Give a critical account of an experiment to determine the thermal conductivity of a material of low thermal conductivity such as cork. Why is it that most cellular materials, such as cotton wool, felt, etc., all have approximately the same thermal conductivity?

One face of a sheet of cork, 3 mm thick, is placed in contact with one face of a sheet of glass 5 mm thick, both sheets being 20 cm square. The outer faces of this square composite sheet are maintained at 100 °C and 20 °C, the glass being at the higher mean temperature. Find (*a*) the temperature of the glass–cork interface, and (*b*) the rate at which heat is conducted across the sheet, neglecting edge effects.

[Thermal conductivity of cork = $6{\cdot}3 \times 10^{-2}$ W m^{-1} K^{-1}, thermal conductivity of glass = $7{\cdot}0 \times 10^{-1}$ W m^{-1} K^{-1}.] (*O. and C.*)

10. (*a*) Discuss the suitability of air as a material for thermal insulation.

(*b*) State the factors which affect the rate of rise of temperature of one end of a metal bar which is heated at the other end.

(*c*) Discuss the relative merits of placing an electrical immersion heater horizontally in a domestic hot-water system either near the top or near the bottom of the tank. (*N.*)

11. Define *thermal conductivity*. Describe briefly the mechanism responsible for the conduction of heat in a non-metallic solid. What is a possible reason for the fact that metals generally have a higher thermal conductivity than non-metals?

A large sheet of rubber rests on the flat surface of the ice on a frozen pond on a day when the air temperature is below 0 °C and the sky is clear. Radiation from the sun falls on the upper surface of the sheet of rubber. Explain what happens to the radiant energy and discuss what factors determine the temperature distribution in the rubber. What conditions are satisfied when the temperature distribution becomes steady? Hence indicate what factors determine whether or not the ice under the sheet of rubber starts to melt. (*O. and C.*)

12. Define *thermal conductivity*. Describe in detail a method of determining the thermal conductivity of cork in the form of a thin sheet.

The base and the vertical walls of an open thin-walled metal tank, filled with water maintained at 35 °C, are lagged with a layer of cork of superficial area 2·00 m^2 and 1·00 cm thick and the water surface is exposed. Heat is supplied electrically to the water at the rate of 250 watts. Find the mass of water that will evaporate per day, if the outside surface of the cork is at 15 °C. [Assume that the thermal conductivity of cork is $5{\cdot}0 \times 10^{-2}$ W m^{-1} K^{-1} and that the specific latent heat of vaporization of water at 35 °C is 2520 kJ kg^{-1}.] (*L.*)

13. Define *thermal conductivity*. Give an account of a method of measuring its value for a metal. An electric heater placed in a small hole at the centre of a metal sphere of diameter 5 cm is supplied with energy at the rate of 100 watts. Find the gradient of temperature (*a*) at a distance of 1 cm from its centre, (*b*) at its outer surface. Assume that the steady state has been attained, and that the thermal conductivity of the metal is 80 W m^{-1} K^{-1}. (*L.*)

14. Define the *thermal conductivity* of a substance. Describe how the thermal conductivity of a metal may be measured, pointing out the sources of error in the experiment.

A large hot-water tank has four steel legs in the form of cylindrical rods 2·5 cm in diameter and 15 cm long. The lower ends of the legs are in good thermal contact with the floor, which is at 20 °C, and their upper ends can be taken to be at the temperature of the water in the tank. The tank and the legs are well lagged so that the only heat lost is through the legs. It is found that 22 watts are needed to maintain the tank at 60 °C. What is the thermal conductivity of steel? When a sheet of asbestos 1·5 mm thick is placed between the lower end of each leg and the floor only 5 watts are needed to maintain the tank at 60 °C. What is the thermal conductivity of asbestos? (*O. and C.*)

15. Describe in detail how you would find the thermal conductivity of a piece of cardboard in the form of a thin disc. Give reasons for the choice of shape of the specimen used in the experiment.

In an experiment using a calorimeter, a quantity of liquid, which is thoroughly stirred, is maintained at a constant temperature by a heater made from a length of resistance wire of uniform circular cross-section covered with a layer of insulation 0·050 cm thick. The potential gradient along the wire is 4·0 V cm^{-1} and the wire is carrying a current of 1·5 A. Calculate the temperature gradient at a point in the insulation 0·025 cm from the surface of the wire. The diameter of the wire is 0·020 cm and the thermal conductivity of the insulating material is 2·0 W m^{-1} K^{-1}. (*N.*)

16. Explain what is meant by the *thermal conductivity* of a metal.

One end of a long uniform metal bar is heated in a steam chest and the other is kept cool by a current of water. Draw sketch graphs to show the variation of temperature along the bar when the steady state has been attained (*a*) when the bar is lagged so that no heat escapes from the sides, (*b*) when the bar is exposed to the air. Explain the shape of the graph in each case.

The surface temperatures of the glass in a window are 20 °C for the side facing the room and 5 °C for the outside. Compare the rate of flow of heat through (i) a window consisting of a single sheet of glass 5·0 mm thick, and (ii) a double-glazed window of the same area consisting of two sheets of glass each 2·5 mm thick separated by a layer of still air 5·0 mm thick. It may be assumed that the steady state has been attained.

[Use the following values of coefficient of thermal conductivity: glass, 1·0 W m^{-1} K^{-1}; air, $2{\cdot}5 \times 10^{-2}$ W m^{-1} K^{-1}.] (*C.*)

17. Define *thermal conductivity* and state a unit in which it is expressed.

Explain why, in an experiment to determine the thermal conductivity of copper using a Searle's bar arrangement, it is necessary (*a*) that the bar should be thick, of uniform cross-section and have its sides well lagged, (*b*) that the temperatures used in the calculation should be the steady values finally registered by the thermometer.

Straight metal bars X and Y of circular section and equal in length are joined end to end. The thermal conductivity of the material of X is twice that of the material of Y, and the uniform diameter of X is twice that of Y. The exposed ends of X and Y are maintained at 100 °C and 0 °C respectively and the sides of the bars are ideally lagged. Ignoring the distortion of the heat flow at the junction, sketch a graph to illustrate how the temperature varies between the ends of the composite bar when conditions are steady. Explain the features of the graph and calculate the steady temperature of the junction. (*N.*)

18. Account as far as you can for the processes involved in the conduction of heat through solids.

Explain why the thermal conductivity of good solid conductors is usually found using the material in the form of bar, whilst that of bad solid conductors is found using a disc.

A metal cylinder of radius R carries a uniform current I. If the resistivity of the material is ρ and its thermal conductivity k, derive an expression for the difference in temperature between the axis and circumference of the cylinder, when steady conditions have been reached. Neglect end effects. (*N.*)

8
Thermal Radiation

We have already seen that "conduction" and "convection" are processes in which the heat is carried from one place to another by the movement of the substance concerned. The heat per day from the sun which reaches the earth, however (an enormous quantity of heat), passes through a considerable region in which there is little or no material substance, and hence heat can pass through a vacuum. In this case we speak of the heat *radiated* by the sun. Radiation *is the name given to the transfer of heat (energy) when the medium takes no part in the transfer*. When we sit in front of a fire the heat reaches us mainly by the process of radiation, since air is a bad conductor of heat and little convection takes place towards us. In this section we shall often use the term "radiation" to denote the heat or thermal energy per second radiated by an object and defer until later the nature of thermal radiation (p. 181).

Detection of heat radiation

The most sensitive method of detecting or measuring thermal radiation is one which utilizes a number of thermocouples in series. A "thermocouple" is the name given to two unlike metals joined together at their ends. When one of the junctions is at a higher temperature than the other, a small electric current flows in the circuit (see p. 8). A *thermopile* is the name given to a series arrangement of thermocouples; fig. 8.1 illustrates a thermopile made of two unlike metals, A, B, such as bismuth and antimony, with a moving-coil mirror galvanometer connected to its terminals. All the junctions *h* are blackened and brought to an open side of a container, while the

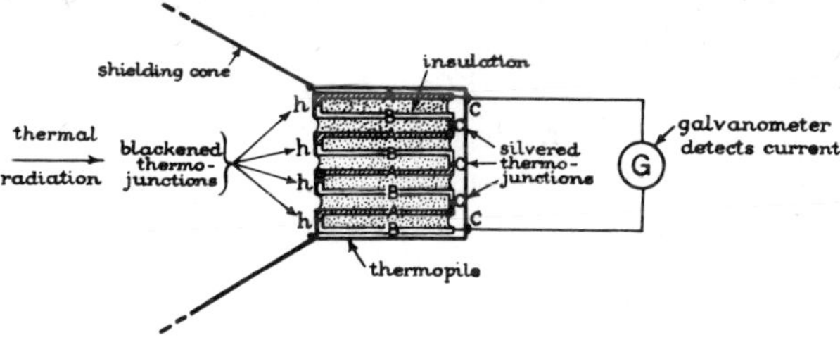

Fig. 8.1 Thermopile (*diagrammatic*)

other junctions *c* are silvered and highly polished and brought to the opposite side. When radiant heat falls on the junctions *h,* they are warmed, and an electric current flows in G, which increases as the amount of radiant heat increases. A cone is placed across the blackened junctions *h* of the thermopile T so as to define the direction of the heat radiation received (fig. 8.1). A *phototransistor* OCP71 is sensitive to heat radiation in a particular band of wavelengths.

Leslie's experiment on total emissive power

In 1804 Sir John Leslie published the results of a research which he called *An Experimental Inquiry into the Nature and Propagation of Heat*. He coated three of the vertical sides of a cube C with a dull black or lampblack surface, a green paint, a white paint, and polished

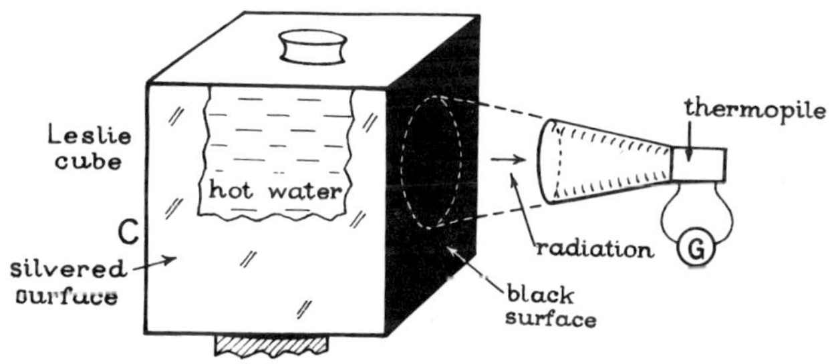

Fig. 8.2 Comparing total emissive powers of surfaces

the fourth side brightly. Hot water was placed inside C and, by turning the cube round, each side was presented in turn, at the same distance, to a detector of radiant heat. Nowadays we should use a thermopile (fig. 8.2). The experiment showed that the radiation from the black surface was greater than that obtained from any other surface, and the radiation from the highly polished surface was least.

By coating the bulb of a sensitive thermometer with lampblack and other materials, Leslie also showed that a dull black surface was an excellent absorber of radiation and that a highly polished, silvered surface was a very poor absorber of radiation. It was thus apparent that *a good absorber of heat is a good radiator of heat, and that a poor absorber of heat is a poor radiator of heat.* This experimental result can also be proved theoretically (p. 191). It may be noted here that we should expect a surface which allows heat to pass through it easily to do so in *either* direction, inward to the body or outward; and hence a good absorber of heat should also be a good radiator. Similarly, a surface which allows little radiation to pass through it inward to the body (a poor absorber) also allows little radiation to pass through it in an outward direction, i.e. it is a poor radiator.

Any radiation incident on a body may be absorbed, reflected or transmitted. An opaque body does not transmit radiation. If most of the incident radiation is absorbed by the body, then very little is reflected. Thus a good absorber of radiation (e.g. a blackened surface) is a poor reflector. Conversely, a poor absorber of radiation (e.g. a polished surface) is a good reflector.

Properties of thermal radiation

At the beginning of the nineteenth century scientists began to investigate the properties and nature of thermal radiation; it was shown that thermal radiation could be reflected from a surface, and refracted into a medium, according to the same laws as light.

The *reflection* of thermal radiation can be demonstrated by placing a hot metal sphere at the focus of a concave mirror A, and positioning another concave mirror B some distance away from A and in front of it. A thermopile placed at the focus of B shows that thermal radiation is received from A, and this could have occurred only by radiation reflected from A and then from B according to the same laws governing the reflection of light. In this experiment the mirror B and the thermopile must be protected, of course, from receiving radiation directly from the hot sphere.

In 1800 Sir William Herschel demonstrated the *refraction* of thermal radiation. He used a prism to produce the visible spectrum of the sun on a screen and found that an appreciable temperature rise occurred when the blackened bulb of a thermometer was placed in the dark region beyond the red. This shows that the thermal radiation from the sun had been refracted by the prism. Figure 8.3 shows the principle of an experiment to show the refraction of thermal radiation.

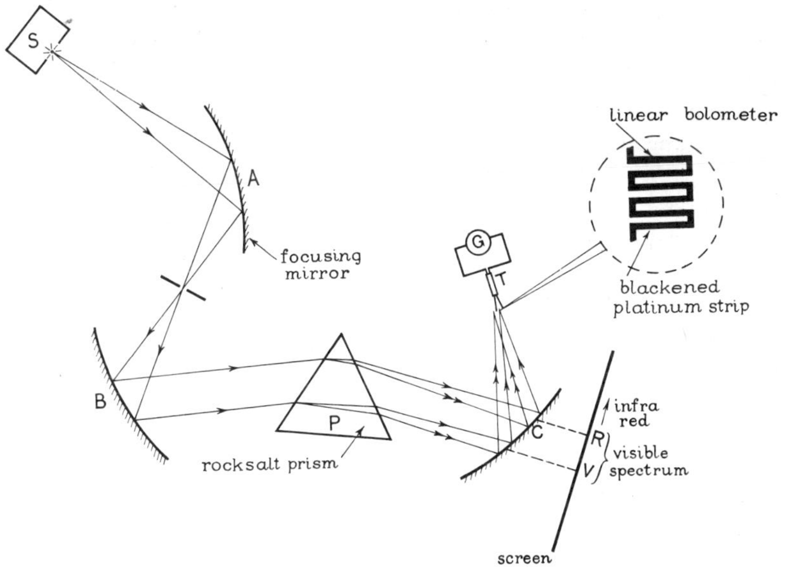

Fig. 8.3 Detection of infra-red rays or thermal radiation

A white-hot source S and two concave mirrors A and B are arranged so that a parallel beam of light is refracted through a rocksalt prism (glass absorbs radiant heat). By means of the concave mirror C, a pure spectrum can be formed as shown. A narrow or linear thermopile T (see p. 186) is used to examine each part of the spectrum. A small deflection is obtained in the thermopile galvanometer as it is moved from the violet to the red end. Beyond the red, in the non-visible region called the *infra-red* region, the deflection increases rapidly before returning to zero further on, showing the presence of thermal radiation.

Experiments also show that thermal radiation exhibits the phenomena of *interference* and *polarization.* Further, light and thermal radiation are both cut off simultaneously at the time of a solar eclipse, and hence their velocity is the same. All experiments show that the nature of thermal radiation and light is the same. Both are types of electromagnetic waves, which we now discuss.

Electromagnetic waves. Infra-red rays

Electromagnetic waves are transverse waves due to electric and magnetic vibrations which travel together. They produce effects according to their frequency of vibration f or wavelength λ. Generally, the speed of any electromagnetic wave in a vacuum $c = f\lambda = 3 \times 10^8$ metres per second (approx).

Radio waves, infra-red rays, visible light, ultra-violet rays, X-rays and gamma rays are all electromagnetic waves. Their wavelengths vary from radio waves, which may have long wavelengths of the order of 1000 m or short wavelengths of the order of a few centimetres, to gamma rays, which have the shortest wavelengths of the order of 10^{-11} m. Visible light has wavelengths in a narrow band, roughly in the range $4{\cdot}0 \times 10^{-7} - 7{\cdot}5 \times 10^{-7}$ m. Ultra-violet rays have shorter wavelengths than $4{\cdot}0 \times 10^{-7}$ m. X-rays have much shorter wavelengths than ultra-violet rays and may approach the wavelengths of γ-rays.

Infra-red rays are invisible rays of wavelengths longer than the red wavelength, $7{\cdot}5 \times 10^{-7}$ m. Most of the heat radiation from hot objects are infra-red rays, roughly in the range $8 \times 10^{-7} - 4 \times 10^{-4}$ m. Radiators at very high temperatures, however, such as some stars, may have a good deal of their heat radiation in the visible spectrum (see p. 188).

Although we experience different sensations from heat radiation and light, the only difference between them is their difference in wavelength (or frequency). Both are forms of energy which travel in space as electromagnetic waves. Glass is *adiathermanous* (opaque) to infra-red rays from fires and may therefore be used as a firescreen. Quartz and rocksalt are diathermanous (transparent) to such rays.

Electroluminescence

Infra-red radiation has been utilized in detection and communication systems, particularly at night. The main feature of a modern system is a gallium arsenide *electroluminescent diode*. This is a semiconductor diode from which radiation is emitted when electrons are injected into the *p-n* junction of the diode by current flow. The radiation consists of photons, which are emitted when the electrons pass to a lower energy level on recombining with the holes at the junction.

The filament of a filament lamp is said to be "incandescent" when current flows. In this case, radiation is due to thermal excitation of atoms and molecules owing to the high filament temperature. In contrast, "luminescence" is due to excitation of electrons and the temperature of the diode remains fairly constant. Further, a wide band of radiation is emitted from a hot filament. In luminescence, however, only a narrow band of wavelengths is emitted. The peak wavelength of the radiation from gallium arsenide is about $0{\cdot}9 \times 10^{-6}$ m at room temperature, which is the near infra-red region.

Infra-red communication system

The principle of an infra-red communication system is shown in fig. 8.4. In the transmitter, the audio-frequency amplifier modulates the infra-red output from the electroluminescent diode Mullard CQY11B. The diode is at the focus of a lens, which emits the radiation as a parallel beam, and this travels to the receiver some distance away. Here the radiation is focused by a lens on to a phototransistor

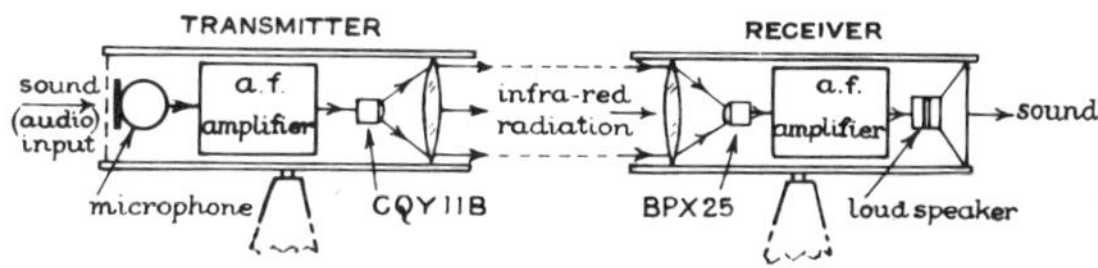

Fig. 8.4 Infra-red communication system

Mullard BPX25. This produces a variation of current proportional to the modulated radiation and the audio-frequency component is then detected and amplified and passed to a loudspeaker.

Further details of the communication system can be obtained from Mullard Educational Service, Mullard House, London, W.C.1.

Prévost's Theory of Exchanges

Suppose that a small hot sphere A, covered with lampblack, is placed inside a larger hollow cold sphere B, similarly coated, so that A is entirely surrounded by B. Experiment shows that, as A becomes colder, B becomes warmer, until eventually A and B are at the same temperature.

The temperature of A begins to fall because heat is radiated from a hot to a cold body (p. 23). It must not be thought, however, that the role of B is a passive one while this happens. According to Prévost's "Theory of Exchanges", *there is a constant stream of radiation from a cold to a hot body*, i.e. from B to A, as well as from A to B. Initially, the amount of heat radiated per unit area by B is less than that emitted per unit area by A. Hence there is a net gain of radiation by B, which therefore rises in temperature. Conversely, the amount of radiation emitted per unit area by A is initially greater than that emitted per unit area by B, and hence the temperature of A decreases. Eventually, the radiation emitted by A and B become equal, and their temperatures are equal. This is an example of "dynamic" equilibrium, in the sense that A and B continue to emit radiant heat but their temperatures are constant.

If A is replaced by a blackened sphere C which is colder than the surrounding sphere B, then B initially emits more radiation to C than C emits to B. The temperature of the latter therefore diminishes while that of C rises. A dynamic equilibrium is eventually obtained between the streams of radiation, when the temperatures of A and B become equal.

Illustrations of the Theory of Exchanges, given above for the case of two bodies at different temperatures, can be extended to any number of bodies inside a constant-temperature enclosure. Eventually all the bodies reach the same temperature. The radiation emitted per unit area from any body is equal to that incident on it from all the other bodies and the walls of the enclosure. Likewise, the radiation emitted per unit area from any part of the walls is equal to that incident on it from all the bodies and the rest of the walls. The radiation or energy per unit volume inside the enclosure is thus constant. Its magnitude depends only on the temperature of the enclosure and not on the properties of the bodies inside it or on the nature of the walls.

Black-body or full radiation

Consider a cavity C with opaque walls and a small hole H in it (fig. 8.5 (i)). If any radiation is incident on H and the reflection factor of the walls is, say, 50%, then after 5 reflections at the walls the reflected radiation is only $(\frac{1}{2})^5$ or about 3% of the incident radiation. Now the chance of any radiation escaping through the small hole

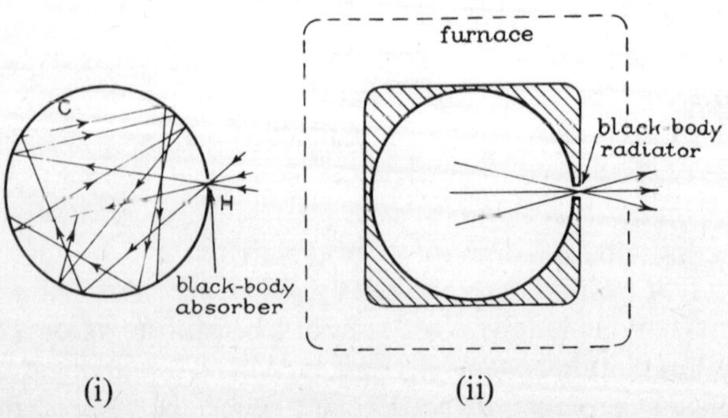

Fig. 8.5 Practical form of black body

after any wall reflection is extremely low. It can therefore be seen that practically all the incident radiation is absorbed by the hole H. This is the case for radiation of all wavelengths incident on H. Thus H acts as a perfect absorber of radiation or as a black body. The hole H appears black when viewed.

A perfect absorber of radiation is a perfect emitter of radiation. Thus if the walls of C are raised to a temperature T, the radiation or energy emitted *from the hole* H is the maximum possible at the temperature T (fig. 8.5 (ii)). All parts of the walls emit (and absorb) the same amount of radiation per unit area in temperature equilibrium, irrespective of the nature of the walls. The radiation from H is thus dependent only on the temperature T. It is called *black-body* or *full radiation*. Note that the black-body radiator is the hole and not the walls of the radiator, and that all wavelengths are emitted at the particular temperature T. Andrade has stated:

> "The glowing heart of a furnace is an ideal (black-body) radiator, for it is practically a small hole surrounded by glowing bodies all at one temperature". When the furnace is cold the heart of the furnace is perfectly black.

In practice, then, a black-body radiator consists basically of a small hole leading to a large cavity of any material, all of which is kept at the same temperature. A small hole in a large metal hollow ball, or a furnace with a small hole in the door, are examples. A hot copper sphere, or the glowing filament of an electric lamp, are not black-body radiators. Certain wavelengths are missing from their spectra. Further, some of the wavelengths present do not contain as much energy as that present in the same wavelengths from a black-body radiator. The radiation from any non-black radiator X will be equal to the radiation of a black-body radiator at a *lower* temperature T, say. We may therefore refer to the "black-body temperature, T" of X, although X is not a black-body radiator.

Investigation of black-body radiation

In 1899 Lummer and Pringsheim carried out an important investigation into the energy obtained from a black-body or full radiator. They used a long narrow porcelain tube coated with black cobalt oxide, and surrounded by two other porcelain tubes, inside a furnace. The two inner tubes were electrically heated by platinum ribbon wound along their entire length, and a porcelain diaphragm, which acted as the radiator, was placed in the middle of the innermost blackened tube. The radiation along the axis of the tube was black-body radiation, and emerged from a hole in the furnace door.

By means of concave mirrors and a fluorite prism, the different wavelengths in the radiation were brought to a focus at different places, so that a pure spectrum was obtained.

A special form of thermopile, called a *linear bolometer,* was used to measure the energy in a narrow band of wavelengths. Basically, it consisted of a blackened narrow platinum strip, which was connected in one arm of a balanced Wheatstone-bridge circuit. When radiation was incident on it, the electrical resistance was changed and the circuit became unbalanced. By passing known currents into the wire the energy required to produce the same change in resistance was found. In this way it was possible to measure the energy in a narrow band, such as 100 Å or 10^{-6} cm, at different wavelengths. The distribution of energy throughout the spectrum of black-body radiation at the particular temperature was thus obtained (fig. 8.3, p. 181).

Distribution of energy in spectrum of black-body radiation

Figure 8.6 illustrates a typical curve for the distribution of energy among the different wavelengths at a particular temperature T such as 1500 K. $E_\lambda \delta\lambda$ is the energy per second per unit area emitted in a narrow band of wavelengths in the interval $\delta\lambda$ from λ to $\lambda + \delta\lambda$. E_λ is called the *emissive power* for the wavelength λ. The unit of E_λ is hence that of "energy per second per unit area per unit wavelength interval", or "watt metre^{-2} per Å or nanometre", for example (W m^{-3}).

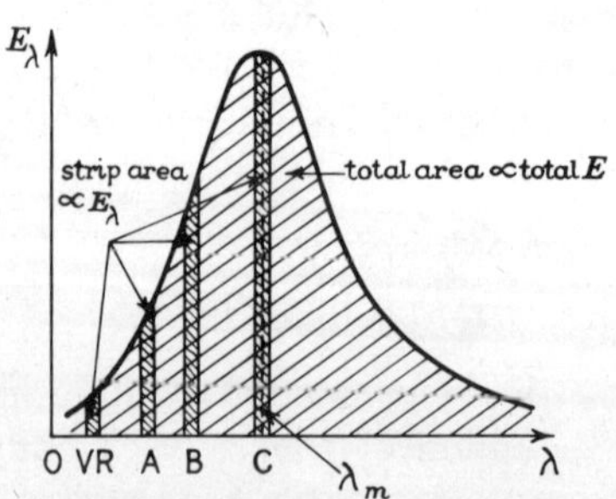

Fig. 8.6 E_λ and E in terms of area

Since $E_\lambda . \delta\lambda$ is the rate of energy per unit area emitted, the energy in a given small band $\delta\lambda$ of wavelengths at A or B is represented on the graph by the shaded areas shown (fig. 8.6). Thus more energy is radiated in the band of wavelengths round B than round A. There is little energy in the very long or very short wavelengths, such as the visible spectrum V – R. The maximum energy occurs in the band of wavelengths at the peak of the curve, corresponding to a wavelength λ_m at C.

It can now be seen that the *total energy* radiated per second per unit area is represented by the area between the curve and the λ-axis. Stefan's law, discussed on p. 188, applies to the total energy radiated.

Planck's law. Wien's law

Lord Rayleigh and Sir James Jeans obtained a formula for E_λ on the basis that the radiation was emitted *continuously* from the black body, but this was not borne out by the experimental results shown in fig. 8.6. In 1900 Planck proposed the then revolutionary principle that the radiation was emitted in definite amounts known as *quanta*, i.e. that the stream of radiation was *discontinuous*. Planck's quantum theory is outside the scope of this book, and we can do no more than quote his result for E_λ, which experiment confirms for all values of λ and T. *Planck's formula* or *law* states that

$$E_\lambda = \frac{a}{\lambda^5(e^{b/\lambda T} - 1)}$$

where a, b are constants and λ, T are respectively the wavelength and absolute temperature of the black body. Thus the energy in a given narrow band of wavelengths depends only on the average wavelength λ of that band and the absolute temperature T of the black body. Figure 8.7 illustrates the variation of E_λ with λ at different temperatures. E_λ is the emissive power at the particular wavelength λ (p. 190).

The wavelength λ_m associated with the maximum energy corresponds to the peaks A, B, C, of the curves and decreases as T increases. At 6000 K, approximately the sun's surface temperature, the maximum energy is obtained in the wavelengths of the visible spectrum. Wien proved theoretically that

$$\lambda_m = \frac{2{\cdot}9 \times 10^{-3}}{T} \text{ metres}$$

where T is in K. This relation is known as *Wien's law*. It shows that $\lambda_m \propto 1/T$. In a qualitative way, we can see that λ_m diminishes as T increases by considering the colour changes when iron is heated

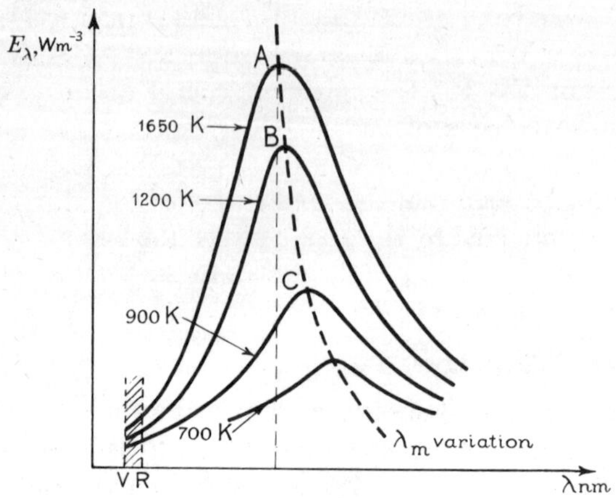

Fig. 8.7 Energy distribution in black body radiation

continuously. The visible colour first observed is dull red, and it then changes through orange to yellow. Very hot stars are "blue", that is, they have an appreciable amount of blue wavelengths in their energy spectrum.

Stefan's law

Planck's and Wien's laws concern the energy in a very narrow band of wavelengths of the radiation emitted from a black body. Stefan made a lucky guess at the law relating the *total* energy from a hot body and its temperature, following a study of some experimental results. He suggested that the total energy per second E emitted from a given black body is proportional *to the fourth power of its absolute temperature* T. Thus $E \propto T^4$, and hence

$$E = \sigma T^4$$

where σ is a constant known as *Stefan's constant*. The fourth-power law was deduced from theoretical arguments by Boltzmann, who showed that it was only true for a black body. It is therefore frequently called the *Stefan–Boltzmann law*. The total energy E is related to E_λ by $E = \int_{\lambda=0}^{\lambda=\infty} E_\lambda \, d\lambda$ (p. 187), and is hence represented by the total area enclosed by the corresponding energy-distribution ($E_\lambda - \lambda$) curve, and the axis of λ. This is shown on p. 186.

Since the SI unit of E is "joules per second per metre²" or "watts per metre²" (W m^{-2}), the SI unit of Stefan's constant is W m^{-2} K^{-4}. The value of σ is about $5{\cdot}67 \times 10^{-8}$ W m^{-2} K^{-4}.

Heat lost by radiation

Suppose a black-body radiator X is initially at an absolute temperature T_1K and is placed inside a black-body enclosure Y at a lower temperature T_0 K. Then the rate of energy emitted per unit area by X is $\sigma T_1{}^4$ and it receives a rate of energy per unit area from Y of $\sigma T_0{}^4$. Thus the net rate of energy emitted per unit area by X is $\sigma T_1{}^4 - \sigma T_0{}^4$ or $\sigma(T_1{}^4 - T_0{}^4)$. Correspondingly, the rate of energy gained per unit area by Y is $\sigma(T_1{}^4 - T_0{}^4)$.

When the absolute temperature T_1 of a black-body radiator is *high* compared with the absolute temperature T_0 of the black-body enclosure, $T_0{}^4$ may be neglected compared with $T_1{}^4$. In this case the rate of heat lost per unit area by the black-body radiator is $\sigma T_1{}^4$ to a good approximation. Thus, if black-body radiators at temperatures of 1000 °C and 1200 °C are placed in turn in a black-body enclosure at a temperature of 15 °C, the respective rates of heat lost per unit area by the hot bodies are practically in the ratio $(1000+273)^4:(1200+273)^4$, or $1273^4 : 1473^4$.

When the absolute temperature T_1 of a black-body radiator is only *slightly higher* than that of the black-body enclosure, the heat lost per second by the radiator is proportional to the *difference in temperature* between itself and the enclosure. Thus suppose T_0 is the enclosure temperature. Then, if $T_1 = (T_0 + x)$, where x is the difference in temperature between T_1 and T_0, the heat lost per second $= \sigma(T_1{}^4 - T_0{}^4) = \sigma[(T_0 + x)^4 - T_0{}^4]$ from Stefan's law. Now $(T_0 + x)^4 - T_0{}^4 = 4T_0{}^3x + 6T_0{}^2x^2 + 4T_0x^3 + x^4$. As T_0 is 300 K for a temperature of 27 °C, for example, and x is assumed small, e.g. less than 30 °C, it can be seen that $6T_0{}^2x^2$, $4T_0x^3$, and x^4 can all be neglected compared with $4T_0{}^3x$. Thus the heat lost per second is $4\sigma T_0{}^3x \propto x$, the temperature difference.

Examples

1. Give an account of Stefan's law of radiation, explaining the character of the radiating body to which it applies and how such a body can be experimentally realized.

If each square centimetre of the sun's surface radiates energy at the rate of 6300 W cm^{-2} and Stefan's constant is $5{\cdot}7 \times 10^{-8}$ W m^{-2} K^{-4}, calculate the temperature of the sun's surface in °C, assuming Stefan's law applies to the radiation. (*L.*)

Energy radiated per second per *metre²* by sun $= 6300 \times 10^4$ W m^{-2}

If T is the absolute temperature of the sun, then

$$\sigma T^4 = 5{\cdot}7 \times 10^{-8}\,T^4 = 6300 \times 10^4$$

$$\therefore\ T = \left[\frac{6300 \times 10^4}{5{\cdot}7 \times 10^{-8}}\right]^{1/4}$$

$$= 5765\ \text{K} = 5492\ ^\circ\text{C}$$

2. Discuss the mechanism by which thermal equilibrium is attained when a hot body is placed in an evacuated space with cooler walls maintained at constant temperature. The body is not in physical contact with the walls.

A diode valve consists of two long coaxial cylinders. The inner cathode cylinder, of radius 0·5 mm, radiates heat like a black body. The radius of the anode cylinder is large compared with that of the cathode. The cathode heater element dissipates one watt per centimetre length. If the steady anode temperature is 227 °C, estimate the temperature of the cathode. Neglect end effects.
[Stefan's constant = $5{\cdot}74 \times 10^{-8}$ W m^{-2} K^{-4}.] (*C.S.*)

For dynamic equilibrium,

energy radiated by cathode per m length = 100 watt.

Since the cathode is effectively in an enclosure of 227 °C or 500 K,

$$\text{energy radiated per m} = \sigma(T^4 - 500^4) \times 2\pi \times 0{\cdot}5 \times 10^{-3}$$

$$= 5{\cdot}74 \times 10^{-8}(T^4 - 500^4) \times 2\pi \times 0{\cdot}5 \times 10^{-3}\ \text{W}$$

$$= 100\ \text{W}$$

$$\therefore\ T^4 - 500^4 = \frac{10^{13}}{5{\cdot}74 \times \pi}$$

$$\therefore\ T = \left[\frac{10^{13}}{5{\cdot}74\pi} + 500^4\right]^{1/4}$$

$$= 886\ \text{K}$$

Emissive power and absorptive power

Total emissive power.—The "total emissive power" e of the surface of a body is defined as the total energy per second per unit area emitted by it. e is thus expressed in watt metre^{-2} (W m^{-2}). Its magnitude depends on the nature and temperature of the surface. The definition of total emissive power given above applies to the surface of any radiating body. In the special case of a black body the total emissive power is denoted by E, and obeys Stefan's law.

The *total emissivity*, ε, of a surface is defined as the ratio of its emissive power e to that E of a black body at the same temperature.

Thus $$e = \varepsilon E = \varepsilon\sigma T^4$$

Emissive power for a wavelength λ.—We have already seen that the radiation from a body contains various wavelengths. The "emissive power of a surface for a given wavelength λ", represented by e_λ, is defined by the statement: "$e_\lambda\delta\lambda$ is the energy per second per unit area emitted in the range of wavelengths λ to $\lambda + \delta\lambda$." Thus, from the definition of total emissive power e, it follows that $e = \int e_\lambda d\lambda$, the integral being taken over the range of wavelengths emitted by the surface of the body. In the special case of black-body radiation, the emissive power for a wavelength λ will be denoted by E_λ, and E_λ depends on the temperature of the body and the wavelength concerned (Planck's law, p. 187).

The *total absorptive power* a of a body is defined as "the fraction of the incident energy absorbed". The *absorptive power for a given wavelength* λ is defined as "the fraction of the incident energy absorbed between the wavelengths λ and $\lambda + \delta\lambda$", and will be denoted by a_λ. Thus $a = \int a_\lambda d\lambda$.

Kirchhoff's law of radiation

In 1833 Ritchie showed by experiment that the ratio of the total emissive power of a surface to its total absorptive power is a constant. In 1860 Kirchhoff showed that this constant was the same for all surfaces, and the following argument can be used to prove this law of radiation.

Consider an object A placed inside a black-body enclosure, e.g. a blackened sphere with a small hole in it (p. 184), and suppose that A and the enclosure are at the same temperature. Then the energy flow per unit area per second inside the enclosure is E, where E depends only on the temperature, and the energy absorbed by the object A is aE per unit area per second, where a is the total absorptive power of A. But in dynamic equilibrium A emits as much energy per unit area per second as it absorbs.

$$\therefore e = aE$$

where e is the total emissive power of A. Thus $a = e/E = \varepsilon$, from above. Hence the total absorptive power is equal to the total emissivity. Also

$$\frac{e}{a} = E \quad \ldots \ldots \ldots \quad (1)$$

Now E is a constant at a given temperature. Hence *the ratio of the total emissive power of a body to its total absorptive power is a constant equal to the total emissive power of a black body at the same temperature.*

The same argument as above can also be used to prove that

$$\frac{e_\lambda}{a_\lambda} = E \quad \ldots \ldots \ldots \quad (2)$$

where e_λ, a_λ are the respective emissive and absorptive powers of the surface for a given wavelength λ, and E_λ is the emissive power for that wavelength of a black body at the same temperature. Thus e_λ/a_λ is a constant, and hence *Kirchhoff's law,* as expression (2) is called, proves that *a good emitter of a certain wavelength is a good absorber of that wavelength.* Experiment shows that the spectrum of a white-hot source of light is crossed by two close dark lines if the light is intercepted by a sodium flame before it is incident on the prism, and that these lines correspond to the wavelengths of the two lines in the light emitted by the sodium flame. This is an example of the statement that a good emitter of a certain wavelength is a good absorber of that wavelength.

From equation (2), $a_\lambda = e_\lambda/E_\lambda = \varepsilon_\lambda$, where ε_λ is defined by the ratio e_λ/E_λ and is called the *emissivity* at a wavelength λ.

Pyrometry measurement of very high temperatures

In Chapter 1 we saw that platinum resistance and gas thermometers are standard instruments which can be used for measuring high temperatures. These instruments cannot be used to measure temperatures above 1500 °C, however, as the bulb containing the wire or gas would melt if it came into contact with the very hot object.

The laws of radiation are used in *pyrometers* for measuring very high temperatures. Pyrometers are used at glass works, at kilns for making bricks, and at steel and iron works. Here manufacturing processes must be regulated at certain known high temperatures.

Total-radiation pyrometer

Féry designed a pyrometer known as a "total-radiation" pyrometer, because it utilizes the total radiation E emitted from a hot object (p. 184). The essential features of the instrument are illustrated in fig. 8.8 (i). C is a highly polished, silvered, concave mirror, which can be moved to or fro along the axis of a tube X by means of a screw D.

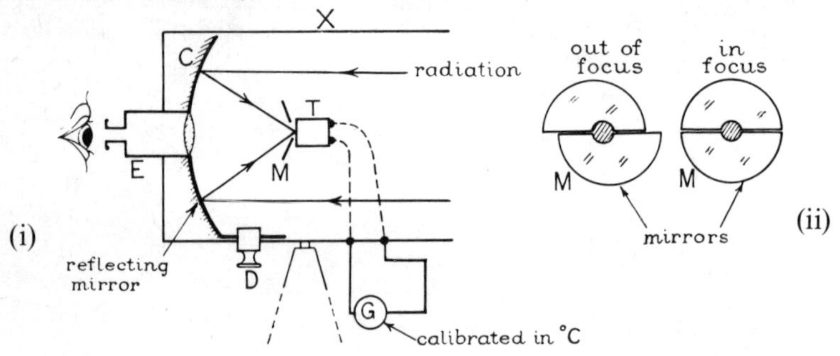

Fig. 8.8 Total radiation pyrometer

The tube is pointed at the furnace whose temperature is required, and after reflection at C the radiation is focused on to the blackened junction of a thermopile T whose other (silvered) junction is shielded from the radiation. The temperature of the furnace is obtained directly from the reading on the millivoltmeter G connected to T, as G is calibrated in degrees Celsius.

Two conditions must be fulfilled before the reading on G is observed. In the first place, the image of the hot body in the concave mirror C must *overlap* the blackened thermopile junction, i.e. the hot body must not be too far away from the pyrometer. In this case, halving the distance of the hot body from C increases the area of its image four times; but the amount of radiation falling on C is four times as much as previously, and hence the energy per unit area per second on the thermojunction is the same as before. The reading on G is thus constant.

In the second place, the radiation from the hot body must be brought to a focus on the blackened junction of T. For this purpose Féry arranged two small semicircular mirrors M near the thermo-junction which are slightly inclined to each other from the vertical (fig. 8.8 (ii)). The centres of the mirrors are cut away so that they form a limiting diaphragm for the radiation received by T. When the mirrors are viewed through the eyepiece E through C, and the rays from the hot body reflected by C are not convergent on to the line of intersection of the two mirrors, the latter are seen displaced, as illustrated in fig. 8.8 (ii). The screw D is then turned so that C is moved, and when the mirrors M are not seen displaced relative to each other, the radiation is focused on to the thermojunction T. The temperature is read directly on an electrical meter joined to the thermopile, previously calibrated in °C by the manufacturer.

Optical pyrometer

Féry's pyrometer utilizes the *total* radiation from the hot object. The *optical pyrometer*, however, utilizes a very narrow band of wavelengths in the radiation from a glowing hot object.

Theory.—The principle of the optical pyrometer can be understood by reference to fig. 8.9, in which X is a glowing black-body radiator and is viewed through a sheet of red glass by an observer.

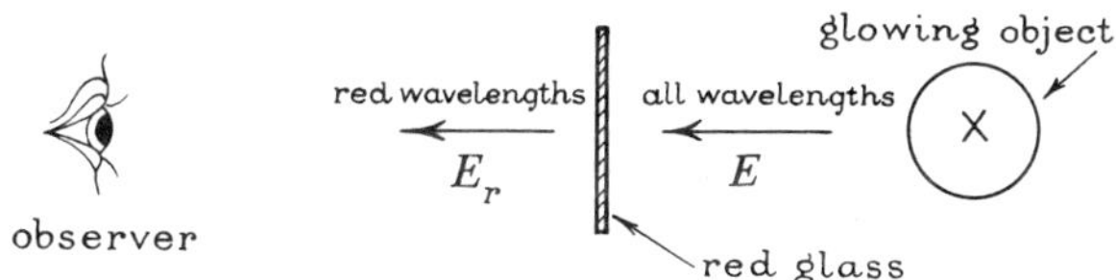

Fig. 8.9 Principle of optical pyrometer

The glass cuts off all wavelengths except a narrow band corresponding to the wavelengths of red light; and hence the *brightness* or intensity E_r of the red light observed is that due to the energy in the red band of wavelengths in the total radiation E from the hot object X. From Planck's law, the brightness of the red light depends only on its wavelength and on the absolute temperature of X (see p. 187).

Suppose now that another glowing black-body radiator D (not shown) is viewed through the same red glass. If the brightness of the red light due to D is the same as that due to X, the temperatures of X and D must be the same, from Planck's law, because the wavelength of the light received by the observer is the same in both cases. Thus, if the temperature of D is known, the temperature of X is also known.

Practical arrangement.—We are now in a position to understand the design of an optical pyrometer which is illustrated in fig. 8.10.

An image I of a glowing furnace D, for example, is formed by the lens L, and is viewed through a piece of red glass R and an eyepiece by an observer O, who thus sees red light due to the radiation from D.

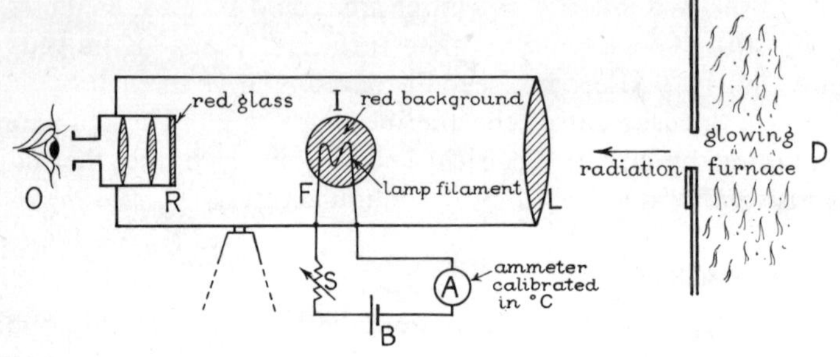

Fig. 8.10 Optical pyrometer

The filament F of an electric lamp is in the same plane as I, and is connected to a 2-volt battery B, a rheostat S, and an ammeter A whose scale is graduated in °C. This has been done by a separate calibration experiment, in which the temperatures of the filament F with various currents in it have been determined by using Planck's law (p. 187).

The current through F controls its temperature. Hence, as the rheostat S is varied, F will appear brighter than the background of the image I of the hot furnace in some cases, and in others it will appear darker than the background of I. For some position of S, the filament F merges into the background and is indistinguishable from it; in this case, since the brightness of the red light from F and from I is the same, the temperatures of F and I are the same. The reading on A is then the temperature of the hot furnace. This type of optical pyrometer is often called a *disappearing-filament pyrometer*, because the filament merges or "disappears" into the background formed by I when the reading on A is taken.

Unlike the total-radiation pyrometer, the optical pyrometer cannot be used to measure the temperatures of hot objects which are not glowing, as wavelengths in the visible spectrum are required for its operation. This is a disadvantage of the optical pyrometer. The optical pyrometer is considered more accurate than the total-radiation pyrometer, however. The theory of an optical pyrometer is relatively less affected by the fact that the hot body may not be a "black body" radiator.

Extension of thermometer scale above 1500 °C

The temperature of a hot object can be found by using a thermocouple which has been previously calibrated by a gas thermometer. Thus the total radiation pyrometer, which utilizes a thermocouple (p. 192), can be used directly to measure temperatures up to the limit of the range of a gas thermometer, which is 1500 °C. This is also the limiting temperature of the filament of a lamp, as the filament would grow thinner by evaporation if the temperature were higher than 1500 °C.

Nevertheless, it is possible to extend the range of pyrometers above 1500 °C by using a disc from which a small sector has been cut away. As an illustration, suppose that this sector has an angle of 20°, and the disc is rotated in front of a very hot furnace at a constant speed of more than 40 revolutions per second. The radiation obtained beyond the disc has now been reduced in the ratio 20° : 360°, i.e. 1 : 18, and we shall suppose that the temperature indicated on a total-radiation pyrometer on which the radiation falls is 1120 °C or 1393 K.

The total radiation from the hot furnace is σT^4, where σ is Stefan's constant (p. 188). Hence the radiation beyond the disc is $\frac{1}{18}\sigma T^4$. But the radiation falling on the pyrometer appears to be that emitted from a body of 1393 K.

$$\therefore \tfrac{1}{18}\sigma T^4 = \sigma 1393^4$$

assuming black-body radiation.

$$\therefore T = (18)^{1/4} \times 1393 = 2870\,\mathrm{K}$$

Thus the temperature of the hot body is 2597 °C.

In general the choice of the size of the angle of the sector depends on the magnitude of the temperature of the hot body, and the method can be applied to obtain a measure of the sun's temperature. A sectored disc can also be used to extend the range of the optical pyrometer beyond 1500 °C, as the brightness of the image beyond the rotating sector is diminished in a ratio depending on the angle of the sector. The calculation for the temperature of the hot body is not as straightforward as that for the case of the total-radiation pyrometer. The brightness of the image of the hot body has also been reduced in a known ratio by using special absorbing glass. Tables are available for reading the temperature in this case.

EXERCISE 8

1. Define a *black-body radiator*. How can such a radiator be realized experimentally? Draw sketches showing roughly how the energy in a black-body radiator varies with the different wavelengths for (i) 700 °C, (ii) 1000 °C, (iii) 1500 °C.

2. State Stefan's law of radiation. A hot metal sphere at 1000 °C cools to 900 °C. If the temperature of the room containing the sphere is constant at 20 °C, compare the rate of cooling of the sphere at these two temperatures, assuming it acts as a black-body radiator. Compare also the rates of cooling when the sphere reaches a temperature of 60 °C and 40 °C respectively.

3. Define *Stefan's constant* and state its units. The temperature of a furnace is 1727 °C. Calculate the heat radiated per cm^2 per minute by the furnace, assuming black-body radiation (Stefan's constant $\sigma = 5{\cdot}7 \times 10^{-8}$ W m^{-2} K^{-4}).

4. The temperature of a furnace, which is not glowing, is 500 °C. Name the pyrometer you would use to determine its temperature, and draw a labelled sketch of its principal features. What other type of thermometer could also be used?

5. Explain the statement: *Stefan's radiation constant is* $5{\cdot}74 \times 10^{-8}$ *W* m^{-2} K^{-4}. Sketch a curve showing the energy distribution in the spectrum of a black-body radiator at a particular temperature. How may the curve be used to obtain a value for Stefan's constant?

The table shows the energy W joules radiated per square centimetre per second from the surface of tungsten at T K.

T	1500	2000	2500	3000	3500
W	5·52	24·0	69·8	160·5	318·0

Assuming that $W = AT^n$, where A and n are constants, find values for A and n and comment on the results obtained. (*L.*)

6. Estimate the filament temperature attained by a 75-W electric lamp, given that Stefan's constant is $5{\cdot}7 \times 10^{-8}$ W m^{-2} K^{-4}. The filament of the lamp has a surface area of 0·80 cm^2 and you may assume that it radiates as a black body. State one further assumption you make in your calculation.

Explain whether the actual temperature would be greater or less than your estimate because of each of these assumptions. (*N.*)

7. Discuss the nature of the processes by which a hot body may lose heat to its surroundings.

A blackened platinum strip of area 0·20 cm^2 is placed at a distance of 200 cm from a white-hot iron sphere of diameter 1·0 cm, so that the radiation causes the temperature, and hence the resistance, of the platinum to increase. It is found that the same increase in resistance can be produced under similar conditions, but in the absence of radiation, when a current of 3·0 mA is passed through the platinum strip, the potential difference between its ends being 24 mV. Estimate the temperature of the iron sphere. (Stefan's constant = $5{\cdot}7 \times 10^{-8}$ W m^{-2} K^{-4}). (*C.*)

8. What do you understand by a *black body*? In what respects does the radiation from a black body at 2000 K differ from that from a black body at 1000 K? How would you devise a black body to radiate at 1000 K?

A blackened sphere, of radius 2·0 cm, is contained within a hollow evacuated enclosure, the walls of which are maintained at 27 °C. Assuming that the sphere radiates like a black body and that Stefan's constant is $5{\cdot}7 \times 10^{-8}$ W m^{-2} K^{-4}, calculate the rate at which the sphere loses heat when its temperature is 227 °C. (*N.*)

9. Explain what is meant by a *black body*. How do the total energy radiated by a black body and its distribution among the wavelengths in the spectrum depend upon the temperature of the radiator?

Describe the structure of an optical pyrometer and explain how it is used to measure the temperature of a furnace. *L.*)

10. How can the temperature of a furnace be determined from observations on the radiation emitted?

Calculate the apparent temperature of the sun from the following information:

Sun's radius: $7{\cdot}04 \times 10^5$ km.
Distance from earth: $14{\cdot}72 \times 10^7$ km.
Solar constant: 0·14 watt per cm^2.
Stefan's constant: $5{\cdot}7 \times 10^{-8}$ W m^{-2} K^{-4}. (*N.*)

11. Give an account of Stefan's law of radiation, explaining the character of the radiating body to which it applies and how such a body can be experimentally realized.

If each square centimetre of the sun's surface radiates energy at the rate of $6{\cdot}3 \times 10^3$ J s^{-1} cm^{-2} and Stefan's constant is $5{\cdot}7 \times 10^{-8}$ W m^{-2} K^{-4}, calculate the temperature of the sun's surface in degrees Celsius, assuming Stefan's law applies to the radiation. (*L.*)

12. What is *black-body radiation*?

Using the same axes sketch graphs, one in each instance, to illustrate the distribution of energy in the spectrum of radiation emanating from (*a*) a black body at 1000 K, (*b*) a black body at 2000 K and (*c*) a source other than a black body at 1000 K. Point out any special features of the graphs.

Indicate briefly how the relative intensities needed to draw one of these graphs could be determined. (*N.*)

13. Explain what is meant by (*a*) a *black body*, (*b*) *black-body radiation.*

State *Stefan's law* and draw a diagram to show how the energy is distributed against wavelength in the spectrum of a black body for two different temperatures. Indicate which temperature is the higher.

A roof measures 20 m × 50 m and is blackened. If the temperature of the sun's surface is 6000 K, Stefan's constant = $5{\cdot}72 \times 10^{-8}$ W m^{-2} K^{-4}, the radius of the sun is $7{\cdot}5 \times 10^{10}$ cm, and the distance of the sun from the earth is $1{\cdot}5 \times 10^{13}$ cm, calculate how much solar energy is incident on the roof per minute, assuming that half is lost in passing through the earth's atmosphere, the roof being normal to the sun's rays. (*O. and C.*)

14. Explain what is meant by *Stefan's constant*, defining any symbols used.

A sphere of radius 2·00 cm with a black surface is cooled and then suspended in a large evacuated enclosure the black walls of which are maintained at 27 °C. If the rate of change of thermal energy of the sphere is 1·848 J s^{-1} when its temperature is −73 °C, calculate a value for *Stefan's constant.* (*N.*)

15. A hot body, such as a wire heated by an electric current, can lose energy to its surroundings by various processes. Outline the nature of each of these processes.

A black body of temperature t is situated in a blackened enclosure maintained at a temperature of 10 °C. When $t = 30$ °C the net rate of loss of energy from the body is equal to 10 watts. What will the rate become when $t = 50$ °C if the energy exchange takes place solely by the process of radiation? What percentage error is there in the answer obtained by basing the solution on Newton's law of cooling? (*C.*)

16. What is Prévost's Theory of Exchanges? Describe some phenomenon of theoretical or practical importance to which it applies.

A metal sphere of 1 cm diameter, whose surface acts as a black body, is placed at the focus of a concave mirror with aperture of diameter 60 cm directed towards the sun. If the solar radiation falling normally on the earth is at the rate of 0·14 watt cm^{-2}, Stefan's constant is 6×10^{-8} W m^{-2} K^{-4} and the mean temperature of the surroundings is 27 °C, calculate the maximum temperature which the sphere could theoretically attain, stating any assumptions you make. (*O. and C.*)

17. Describe a black body suitable for a laboratory investigation of the properties of black-body radiation.

In one diagram draw spectral curves (emissive power against wavelength) for black-body radiation from a source at (*a*) a very high temperature, e.g. that of the sun (*c*. 6000 K), (*b*) the temperature of melting platinum (*c*. 2000 K), (*c*) the temperature of boiling water (*c*. 400 K). Mark on the wavelength axis the range covered by the visible spectrum. State how such curves illustrate (i) Stefan's law, (ii) a relationship between the absolute temperature of the source and the wavelength at which the emissive power at that temperature is a maximum.

A conducting sphere, diameter 1 cm, whose surface has an emissivity coefficient of 0·1, hangs at the end of a copper wire 1 m long and 1 mm in diameter in an evacuated thermally conducting box, which is very large compared with the sphere and its support. The temperature of the sphere is 100 °C and that of the end of the wire where it joins the box is 20 °C. Find the ratio of the rates at which the sphere initially loses heat by radiation and conduction respectively. (Thermal conductivity of copper = 380 W m^{-1} K^{-1}; Stefan's constant = $5{\cdot}75 \times 10^{-8}$ W m^{-2} K^{-4}.) (*O. and C.*)

9

Introduction to Thermodynamics

Thermodynamics is the study of the conversion of energy from one form to another. In its widest form the subject embraces the energy changes in an electric cell, or a stretched wire, or a chemical reaction, for example, but we shall consider only the most common change of energy, between *heat* and *work* or *mechanical energy*.

First law of thermodynamics

The concept of energy grew gradually from experiments and ideas, notably by Joule, Meyer, Kelvin, Clausius and others from about 1840 onwards. By a series of careful experiments Joule was able to show that the temperature of a substance could be raised by doing mechanical work on it, and that there was the *same constant ratio* between the work expended and the heat produced. His experiments were carried out with a wide range of materials and his methods included those of electrical heating. The conclusion was that mechanical work and heat are equivalent (p. 13). It is convenient to regard them as forms of *energy*.

Machines are designed to produce energy in one form at the expense of an equivalent amount of energy of another form. An electric motor, for example, converts electrical energy mainly to mechanical energy. Some heat and sound energy are also produced. It was realized gradually that the total energy obtained from or stored in a system must be equal to the total energy supplied to it, although the energy may be transformed from one form to another. This is known as the *Principle of the Conservation of Energy* or the *First Law of Thermodynamics*. It implies that energy can neither be created nor destroyed, and that "perpetual motion" machines cannot run for ever without the expenditure of some energy on them, even though this may be very small.

Conversion of work and heat

When the rim of a rotating metal wheel rubs against a metal surface, the frictional forces produce heat. This is a conversion of mechanical energy to heat. It takes place with nearly 100% efficiency, since all the work done against friction is practically all converted into heat.

An important example of the reverse phenomenon, the conversion of heat to mechanical energy, occurs when a gas in a cylinder is heated. If the gas is allowed to expand, some of the heat supplied is used for doing work against the external pressure to which the gas was subjected. This is called *external work*. Assuming no friction, the rest of the heat supplied is used for increasing the kinetic energy of the gas molecules as the volume increases. The rest of the heat is thus spent in increasing the *internal energy* of the gas. Hence, from the First Law of Thermodynamics,

$$\delta Q = \delta U + \delta W$$

where δQ is the amount of heat supplied, δU is the consequent increase in internal energy and δW is the external work done by the gas (p. 100).

As we have seen, $\delta W = p\,.\,\delta V$ under reversible conditions, where p is the pressure and δV is the increase in volume of the gas (p. 109). The change δU in internal energy can also be calculated. When a gas is heated at constant volume, the increase in internal energy is equal to the heat supplied, because no external work is done in this case. If the mass of the gas is m, the principal specific heat capacity of the gas at constant volume is c_V and the temperature rise is δT, then $\delta Q = m\,.\,c_V\,.\,\delta T$. Thus $\delta U = m\,.\,c_V\,.\,\delta T$. The increase in internal energy per unit mass of gas is $c_V\,.\,\delta T$. As the attraction between the molecules of an ideal or perfect gas is negligible, the internal energy of this gas is independent of its volume. Consequently any change δT in the temperature of the gas, for example, in an adiabatic expansion, produces a change in the internal energy equal to $c_V\,.\,\delta T$ per gram. Thus, since $\delta Q = \delta U + p\,.\,\delta V$ for an ideal gas under reversible conditions,

$$\delta Q = m\,.\,c_V\,.\,\delta T + p\,.\,\delta V$$

Not all changes in internal energy are associated with temperature change. In latent heat of vaporization, for example, some of the latent heat supplied at the boiling-point increases the internal energy of the substance concerned without temperature change, by changing it from liquid to vapour against the inter-molecular attraction. The remainder of the latent heat is used for doing work against the external atmospheric pressure when the liquid expands in changing to vapour, and is calculated from the relation:

external work $=$ *pressure* $\times$ *volume change* (see p. 34)

Conversion of heat to mechanical work

We have already noted that rubbing one metal surface against another converts nearly all the work done against friction into heat and that the conversion is thus almost 100% efficient. As we shall now show, however, a practical *machine* or *engine* which converts heat to mechanical work continuously can never be almost 100% efficient.

A substance such as a gas used in an engine must clearly be brought back to its initial state after a series of operations· so that the changes can be repeated indefinitely and the engine can function continuously. The gas must therefore be taken through a *cycle* of

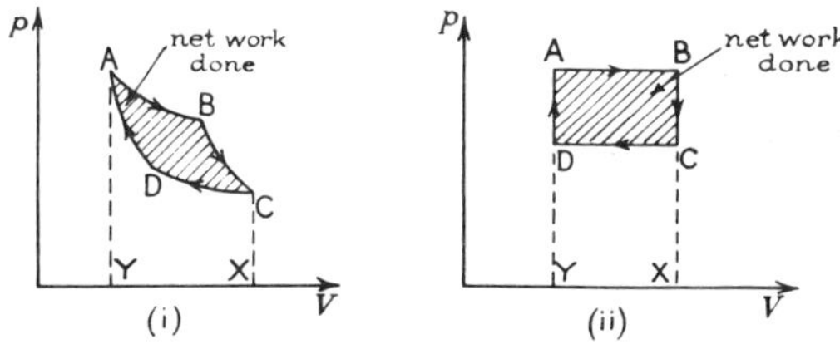

Fig. 9.1 Conversion of heat to work

pressure-volume (p-V) changes. An example of a cycle is shown by ABCDA in fig. 9.1 (i) and (ii). From A to B, and from B to C, the gas does mechanical work equal to the integral of $p\,.\,dV$ between the limits concerned, and this is represented by the area ABCXYA between the curve ABC and the volume axis (see p. 99). Some heat Q_1 must be taken in by the gas to do this work. The gas can only be restored to its original state, represented by A, by compressing it along CD and then DA. In this case work is done *on* the gas represented by the area CDAYXC, and the gas will then give up an amount of heat Q_2 which is less than Q_1. Since there is no change in the internal energy of the gas at the end of the cycle, it follows from the law of conservation of energy that the difference in heat $Q_1 - Q_2$ is equal to the net mechanical work done *by* the gas, which is represented by the area ABCD.

The *efficiency* E of the engine is defined by

$$E = \frac{\text{work done}}{\text{heat supplied}} \times 100\,\% = \frac{Q_1 - Q_2}{Q_1} \times 100\,\%$$

$$= \left(1 - \frac{Q_2}{Q_1}\right) \times 100\,\%$$

It therefore follows that a heat engine can never be 100% efficient. An engine can never convert all the heat energy supplied to it into mechanical energy.

Carnot cycle

Sadi Carnot, in 1824, saw that no engine could be expected to have a higher efficiency than one working under reversible conditions, which is an ideal case. He chose to consider a reversible cycle of isothermal and adiabatic changes as an example, not the only one, of the maximum efficiency that could be expected, and this is known as a *Carnot cycle*. Although we shall apply the Carnot cycle to the case of a gas in an engine, the Carnot cycle of operations can be performed on any other thermodynamic system, such as a liquid surface under surface tension forces or electrons in a thermocouple.

The stages are shown in fig. 9.2 (i).

(1) First, a gas is placed in thermal contact with a large reservoir or source always at a constant temperature T_1 and expands isothermally and reversibly from a state represented by A in fig. 9.2 (ii)

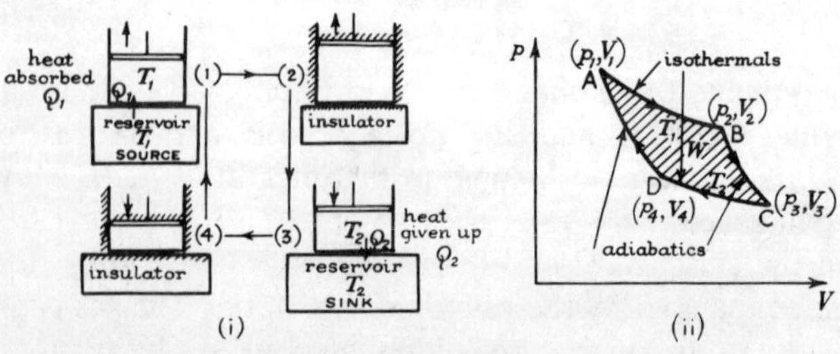

Fig. 9.2 Carnot cycle stages

to one represented by B. Here the gas absorbs a quantity of heat Q_1 from the source, and as conditions are reversible, an equivalent amount of external work is done by the gas.

(2) The gas, now insulated, expands adiabatically and reversibly from a state represented by B in fig. 9.2 (ii) to that represented by C. Since no heat enters or leaves the system, the energy for the work done by the gas is taken from its internal energy, thus lowering its temperature to T_2.

(3) The gas is now placed in thermal contact with a large reservoir or "sink" always at a constant temperature T_2, and it is isothermally and reversibly compressed from C to D. Work is then done *on* the gas, and an equivalent amount of heat Q_2 is rejected to the "sink"

(4) Lastly, the gas is compressed adiabatically and reversibly from D, and returns to A. This restores the gas to its original state. The work done on the gas serves to increase its internal energy back to its original value.

On balance, therefore, the external work done on the gas, $W = Q_1 - Q_2$, and this is represented by the area ABCD. In the Carnot cycle, note that *all* the heat absorbed is taken from a reservoir at a constant high temperature T_1 and that *all* the heat rejected is given back to a reservoir at a constant lower temperature T_2. The *efficiency E* of the cycle is given by

$$E = \frac{W}{Q_1} = \frac{Q_1 - Q_2}{Q_1} = 1 - \frac{Q_2}{Q_1}$$

and to calculate E we need to consider the details of the isothermal and adiabatic changes. This will be left to later (p. 207). First, however, we shall show that the Carnot cycle itself is of special importance in consideration of its efficiency.

Carnot refrigerator. Efficiency of reversible engines

As we have just seen, the essential features of an engine are that, in a cycle, heat Q_1 is extracted from a source at a high temperature T_1 and a smaller amount of heat Q_2 is given up to a "sink" at a lower temperature T_2. The difference $Q_1 - Q_2$ is delivered as mechanical

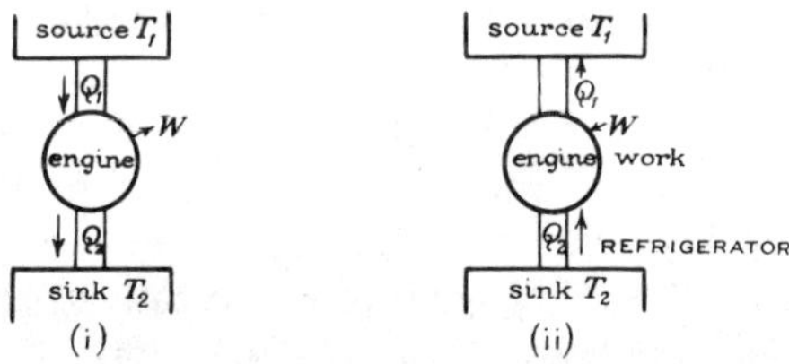

Fig. 9.3 Engine and refrigerator

work W (fig. 9.3 (i)). If the engine can be reversed, it will function as a *refrigerator*. In a cycle, a quantity of heat Q_2 is now extracted from a cold body at T_2, and an amount of heat Q_1, equal to $(Q_2 + W)$, is delivered to a hotter reservoir at T_1 (fig. 9.3(ii)). The corresponding pressure-volume changes are traced out in a reverse way to that shown in fig. 9.2 (ii). If all the processes are reversible, the magnitudes of Q_1, Q_2 and W are exactly the same as when the engine is run in reverse, but the directions of heat flow are reversed (compare fig. 9.3 (i) and (ii)).

Suppose now that an engine A is arranged to drive a *reversible* (e.g. Carnot cycle) engine R as a refrigerator, that is, in reverse, by means of the work W which A delivers to R (fig. 9.4). The efficiencies of the two engines are:

$$E_A = \frac{W}{Q_1{}^A} \qquad E_R = \frac{W}{Q_1{}^R}$$

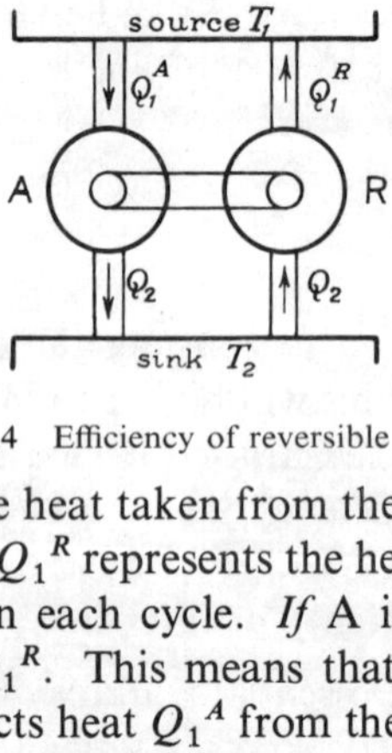

Fig. 9.4 Efficiency of reversible engine

where $Q_1{}^A$ represents the heat taken from the source at a temperature T_1 by the engine A, and $Q_1{}^R$ represents the heat given up to the source by the refrigerator R, in each cycle. *If* A is more efficient than R, then $Q_1{}^A$ is *less* than $Q_1{}^R$. This means that, over a complete cycle, the whole system abstracts heat $Q_1{}^A$ from the source and returns to it an amount of heat $Q_1{}^R$ which is more than that extracted. Now no external work is done by the system, as the whole of the work supplied by A has been used to drive R. Consequently the net effect is to extract heat from the sink and deliver it to the source at a higher temperature, without any external agency being used. Thus we could afford to run R so that it only delivered heat $Q_1{}^A$ to the source in each cycle of A, and use the spare external work for some other purpose. This work would be extracted from the "sink", leaving the source unaffected. The effect would be to create a perpetual-motion machine, which converts heat from a sink into work, without requiring any other change or external agency; this would be similar to a ship driving itself by taking in heat from the ocean. This process is contrary to experience. Hence the engine A *cannot* have a greater efficiency than a reversible engine, that is, *a reversible engine has a maximum efficiency*. We also deduce that all reversible engines, operating between a source and a sink at the same temperatures in each case, have the same efficiency.

Second law of thermodynamics

The *Second Law of Thermodynamics* is a precise general statement of experience to which we have just referred. It has several forms:

1. (Clausius). It is impossible for any self-acting machine working in a cyclical process unaided by an external agency to make heat pass from one body to another at a higher temperature.
2. (Kelvin). It is impossible by means of inanimate material agency to derive mechanical effect from any portion of matter by cooling it below the temperature of the coldest body of its surroundings.
3. (Planck). It is impossible to construct an engine which, working in a complete cycle, will produce no effect other than the raising of a weight (i.e. mechanical work) and the cooling of a heat reservoir.

These three statements can be shown to be equivalent to each other.

A refrigerator extracts heat from a body and passes it to a warmer body· but it is not a self-acting machine. It is driven by an electric motor, for example, so there is no violation of the second law of thermodynamics. In principle, it would be possible to convert some of the internal energy of the ocean to mechanical energy or work. This would not violate the first law of thermodynamics. But it would be impossible for a ship to be driven continuously by this process, because the second law would be violated. Here there is only one reservoir, and work can only be done in a cyclical process when heat is abstracted from a reservoir at a high temperature and rejected to one at a lower temperature.

Thermodynamic temperature scale

We have seen that the efficiencies of all perfectly reversible engines working between the same two temperatures are exactly the same. The efficiency depends only on these temperatures and not on the nature of the working substance in the engine, which might be a gas or a mixture of liquid and vapour, for example. This led Lord Kelvin to suggest a *thermodynamic temperature scale* or *Kelvin temperature scale* (K). This has the merit of being independent of the nature of a substance, whereas the temperature on the gas thermometer, or platinum resistance thermometer, or mercury-in-glass thermometer scales depends on the respective properties of a gas, platinum, and mercury-in-glass.

On the thermodynamic or Kelvin scale, the ratio of two temperatures is defined by the relation

$$\frac{T_1}{T_2} = \frac{Q_1}{Q_2} \quad . \quad . \quad . \quad . \quad . \quad . \quad . \quad \text{(i)}$$

where T_1, T_2 are the respective temperatures of the source and sink of a reversible engine and Q_1, Q_2 are the respective quantities of heat extracted from the source and given up to the sink. Now the efficiency, E, is given by

$$E = \frac{W}{Q_1} = \frac{Q_1 - Q_2}{Q_1} = 1 - \frac{Q_2}{Q_1} = 1 - \frac{T_2}{T_1}, \text{ by definition.}$$

$$\therefore E = \frac{T_1 - T_2}{T_1} \quad . \quad . \quad . \quad . \quad . \quad . \quad \text{(ii)}$$

Thus the efficiency of a reversible engine, which depends only on the temperature of its source and sink, can be used to define the thermodynamic scale of temperature.

In order to put the thermodynamic scale on a numerical basis, the fixed point known as the *triple point of water* is given the value of 273·16 K; this is the temperature at which ice, water and water vapour coexist in equilibrium (see p. 143). Thus if a Carnot engine operates between a source at a thermodynamic temperature T and a sink at a thermodynamic temperature T_0, the triple point of water, then the heats Q and Q_0 absorbed and rejected are given by

$$\frac{T}{T_0} = \frac{Q}{Q_0}$$

or

$$T = T_0 \frac{Q}{Q_0} = \left(273{\cdot}16 \times \frac{Q}{Q_0}\right) \text{K}$$

As Q tends to zero, T tends to zero. The efficiency E of the engine, which is given by $E = 1 - (Q/Q_0)$, then becomes 1 when the absolute zero is reached, that is, when the sink and source are respectively at absolute zero and slightly above absolute zero. This provides a definition of absolute zero on the thermodynamic temperature scale, although it is impossible to achieve in practice, and it is independent of the properties of the particular substance used in the Carnot cycle.

Thermodynamic and ideal gas temperature scales

We shall now show that the thermodynamic or Kelvin temperature scale is identical with the ideal gas temperature scale. To do this, suppose an ideal gas is the working substance taken round the Carnot cycle represented by ABCD in fig. 9.5.

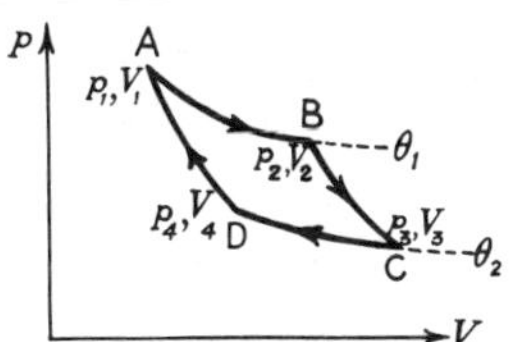

Fig. 9.5 Thermodynamic and ideal gas temperatures

Along the isothermal AB, suppose a quantity of heat Q_1 is taken in from the reservoir, and the state of the gas passes from A, where it has a pressure p_1, volume V_1 and absolute temperature θ_1 on the *ideal gas scale*, to B, where its state is represented by p_2, V_2, θ_1. Along the adiabatic BC, suppose the pressure, volume and absolute temperature change to p_3, V_3, θ_2 at C, and then to p_4, V_4. θ_2 when the gas is compressed isothermally to D, giving up a quantity of heat Q_2 to the sink. The gas returns to its original state at A when it is compressed adiabatically from B to A.

Since the gas expands isothermally and reversibly from A to B, the work done is equal to the heat Q, obtained from the reservoir. Thus, from p. 109,

$$Q_1 = \int_{V_1}^{V_2} p \,.\, dV = R\theta_1 \log_e \left(\frac{V_2}{V_1}\right) \qquad . \quad . \quad . \quad . \quad \text{(i)}$$

where R is the gas constant for the given mass of gas. Similarly, the heat Q_2 returned to the sink is given by

$$Q_2 = \int_{V_4}^{V_3} p \,.\, dV = R\theta_2 \log_e \left(\frac{V_3}{V_4}\right) \qquad . \quad . \quad . \quad . \quad \text{(ii)}$$

From the isothermals AB, CD,

$$p_1V_1 = p_2V_2 \quad \text{and} \quad p_4V_4 = p_3V_3$$

From the adiabatics BC, DA,

$$p_2V_2^{\gamma} = p_3V_3^{\gamma} \quad \text{and} \quad p_4V_4^{\gamma} = p_1V_1^{\gamma}$$

$$\therefore p_2V_2 \times V_2^{\gamma-1} = p_3V_3 \times V_3^{\gamma-1}$$

and $$p_1V_1 \times V_1^{\gamma-1} = p_4V_4 \times V_4^{\gamma-1}$$

Dividing and using $p_1V_1 = p_2V_2$ and $p_4V_4 = p_3V_3$, we obtain

$$\left(\frac{V_2}{V_1}\right)^{\gamma-1} = \left(\frac{V_3}{V_4}\right)^{\gamma-1}$$

or $$\frac{V_2}{V_1} = \frac{V_3}{V_4}$$

Hence, from (i) and (ii), $$\frac{Q_1}{Q_2} = \frac{\theta_1}{\theta_2} \quad \text{(iii)}$$

But $$\frac{Q_1}{Q_2} = \frac{T_1}{T_2}$$

by definition of the thermodynamic temperature scale (p. 206).

Hence $$\frac{T_1}{T_2} = \frac{\theta_1}{\theta_2} \quad \text{(iv)}$$

When θ_1 is zero, T_1 is zero. Hence the absolute zero on the ideal gas scale is the same as that on the thermodynamic or Kelvin scale. If the triple point of water has the same numerical value, 273·16, on each scale, it follows from (iv) that any other temperature is identical on the two scales. Thus although the thermodynamic or Kelvin scale of temperature originated from considerations of reversible engines, it is realized in practice by measuring temperature with a gas thermometer, and then correcting the result to the ideal gas thermometer scale by using the equation of state of the gas (p. 142).

Example

Distinguish between *isothermal* and *adiabatic* changes.

A constant mass of ideal gas is taken round a cycle represented by ABCD on a pressure-volume diagram. AB and DC are isothermals at 150 °C and 27 °C respectively, and AD and BC are adiabatics. If the volumes at A and B are respectively 20 litres and 50 litres and the pressure at A is 10 atmospheres, calculate (*a*) the heat entry along AB, (*b*) the net work done by the gas during the complete cycle.

Assume 1 atmosphere = 10^5 N m^{-2}. Take $\log_e 10 = 2{\cdot}30$. (*N.*)

(*a*) The heat entry along AB (fig. 9.6) = the external work done by the gas

$$= RT_1 \log_e\left(\frac{V_2}{V_1}\right) \text{ (see p. 207)}$$

$$= p_1V_1 \log_e\left(\frac{V_2}{V_1}\right) \quad \text{(i)}$$

But $p_1 = 10 \times 10^5$ N m^{-2}, $V_1 = 20 \times 10^{-3}$ m^3, $V_2 = 50 \times 10^{-3}$ m^3

$$\therefore \text{ heat } Q_1 = 10^6 \times 2 \times 10^{-2} \log_e (\tfrac{5}{2}) \text{ J}$$
$$= 2 \times 10^4 \times 2{\cdot}30 \log_{10} (\tfrac{5}{2}) \text{ J}$$
$$= 2 \times 10^4 \times 2{\cdot}30 \times 0{\cdot}3980 \text{ J}$$
$$= 18\,300 \text{ J}$$

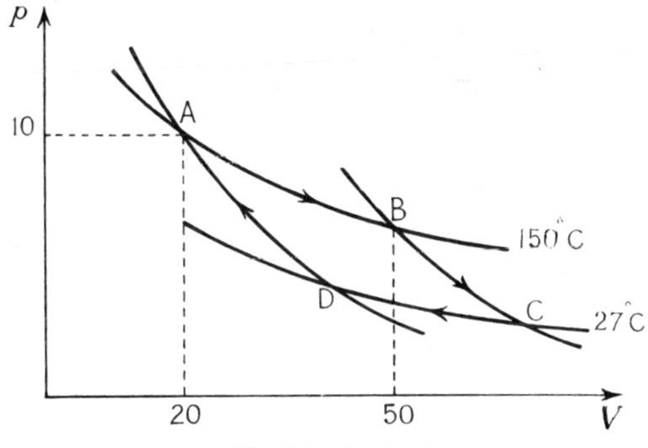

Fig. 9.6 Example

(*b*) The net work done by the gas during the cycle ABCD (fig. 9.6)

$$= \text{total heat absorbed} = Q_1 - Q_2$$

where Q_1 is the heat absorbed along AB and Q_2 is the heat given up when the gas is compressed from C to D. No heat change occurs along BC or DA as these are adiabatics.

We have already calculated Q_1. Similarly,

$$Q_2 = RT_2 \log_e \left(\frac{V_3}{V_4}\right) \quad \ldots \ldots \quad \text{(ii)}$$

where V_3 is the volume at C (pressure p_3 say),

and V_4 the volume at D (pressure p_4 say).

Now, from p. 208,

$$\therefore \frac{V_3}{V_4} = \frac{V_2}{V_1} = \frac{50}{20} = \frac{5}{2} \quad \ldots \ldots \quad \text{(iii)}$$

Also, $$p_1 V_1 = RT_1 \quad \text{or} \quad R = \frac{p_1 V_1}{T_1} \quad \ldots \ldots \quad \text{(iv)}$$

Substituting for R and V_3/V_4 in (ii),

$$\therefore Q_2 = \frac{p_1 V_1 T_2}{T_1} \log_e \left(\frac{5}{2}\right)$$

Hence, with (i), $$Q_1 - Q_2 = p_1 V_1 \log_e \left(\frac{5}{2}\right) - \frac{p_1 V_1 T_2}{T_1} \log_e \left(\frac{5}{2}\right)$$

$$= p_1 V_1 \log_e \left(\frac{5}{2}\right) \left[1 - \frac{T_2}{T_1}\right]$$

$$= Q_1 \left[1 - \frac{T_2}{T_1}\right] = 18\,300 \times \left(1 - \frac{300}{423}\right) \text{ J}$$

$$= 5320 \text{ J}$$

Refrigerator or heat pump

As we saw on p. 204, it is sometimes possible to drive a reversible Carnot cycle engine "backwards", so that heat is transferred from the cold body or sink to the hotter body or source by a motor doing

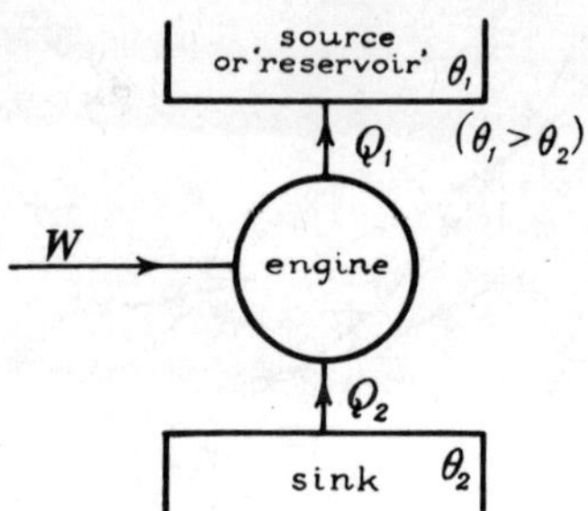

Fig. 9.7 Heat pumped from sink to source or reservoir

work W. This is illustrated in fig. 9.7. In this case we have a *refrigerator* or a *heat pump,* a device for extracting or delivering heat from a cold body.

For a refrigerator, the *coefficient of performance* is defined as

$$\frac{\text{heat extracted from sink}}{\text{work done}} = \frac{Q_2}{W}$$

From the first law of thermodynamics, $W = Q_1 - Q_2$.

$$\therefore \text{coefficient of performance} = \frac{Q_2}{Q_1 - Q_2} = \frac{T_2}{T_1 - T_2} \quad . \quad (1)$$

For a heat pump, the coefficient of performance is defined as

$$\frac{\text{heat transferred to source}}{\text{work done}} = \frac{Q_1}{W}$$

$$= \frac{Q_1}{Q_1 - Q_2} = \frac{T_1}{T_1 - T_2} \quad . \; . \; . \; . \; . \; . \; . \quad (2)$$

From (1) or (2) the efficiency increases when the difference in temperature between the source and sink is small, i.e. only a small amount of mechanical energy W is needed to extract a large amount of heat this case.

Example

A Carnot engine uses an ideal gas as a working substance. It is driven in reverse by a $\frac{1}{2}$-kW electric motor of efficiency 60% in order to freeze water at 0 °C. Assuming the working temperatures of the engine are 15 °C and 0 °C, and no heat losses in the refrigerator unit, find the time taken to freeze 10 kg of water. (Specific latent heat of fusion of ice $l = 336$ kJ kg^{-1}.)

Power output of motor = 60% of 500 watts = 300 watts.

From theory of refrigerator, heat Q_2 abstracted is related to work done W by

$$\frac{Q_2}{W} = \frac{Q_2}{Q_1 - Q_2} = \frac{T_2}{T_1 - T_2} = \frac{273}{15}$$

$$\therefore Q_2 = \frac{273}{15} \times W$$

$$\therefore \text{heat abstracted per second} = \frac{273}{15} \times 300 = 5460 \text{ watts.}$$

To freeze 10 kg of water at 0 °C, heat extracted = 10 × 336 kJ = 3360 × 10^3 J

$$\therefore \text{time required} = \frac{3360 \times 10^3}{5460} \text{ seconds}$$

$$= \frac{3360 \times 10^3}{5460 \times 60} \text{ min} = 10{\cdot}3 \text{ min.}$$

EXERCISE 9

1. What is the first law of thermodynamics? When a fixed mass of gas is compressed, what quantitative relation does the law establish between the external work done, the internal energy of the gas, and the heat which leaves the system? When can the conditions of the compression be described as (*a*) isothermal and (*b*) adiabatic? Show how the law leads to the relation pV^{γ} = constant for reversible adiabatic changes in an ideal gas.

Explain qualitatively why the work done in the reversible compression of a given mass of an ideal gas from an initial volume V_1 to a final volume V_2 is greater if the change takes place adiabatically than if it takes place isothermally, the initial temperature in both cases being the same. (*O.* and *C.*)

2. What is meant by (*a*) an isothermal change, (*b*) an adiabatic change in a gas? Explain how you would attempt to achieve each type of change experimentally.

Three litres of an ideal gas at atmospheric pressure and 27 °C are compressed adiabatically to a volume of 1 litre. Calculate the resultant pressure. At its new temperature the gas is expanded isothermally to its original volume. Calculate the final pressure and temperature. Assume that the ratio of the specific heat capacities of the gas is 1·67.

Illustrate the whole process on a *p*, *v* diagram. By referring to the diagram, explain which is greater—the work done on the gas, or the work done by the gas. (*N.*)

3. What do you understand by the *internal energy* of a system?

A rigid metal vessel initially contains a mixture of hydrogen and oxygen at room temperature. The mixture is ignited by a spark (which itself produces negligible heat) and the temperature of the vessel and its contents is allowed to return to room temperature. Has the internal energy of the contents of the vessel increased or decreased? Give reasons for your answer and explain briefly how you would measure the change.

A certain mass of a perfect gas undergoes a change from an initial pressure P_1 and volume V_1 to a final pressure P_2 and volume V_2 such that $P_2V_2 > P_1V_1$. Find the change in internal energy of the gas by considering that the change is effected in two stages: an initial heating at constant volume followed by further heating at constant pressure. (*O.* and *C.*)

4. What do you understand by the *internal energy* of a gas?

53 g of an ideal gas initially at a pressure of 32 atmospheres and at 27 °C occupy a volume of 1 litre. If the gas is allowed to expand reversibly and adiabatically to a volume of 8 litres, calculate (*a*) the final pressure, (*b*) the final temperature, (*c*) the work done by the gas during the expansion.

The gas in its initial state is now taken to the same final state in two stages, namely: (i) at constant pressure until the volume is 8 litres, followed by (ii) cooling at constant volume. Illustrate these changes on a p, V diagram and calculate the temperature at the end of stage (i). Also discuss the energy changes which occur during the two stages. (Specific heat capacity at constant volume and constant pressure = 0·30, 0·50 kJ kg^{-1} K^{-1} respectively; 1 atmosphere pressure = 10^5 N m^{-2}). (*N.*)

5. What is meant by an *adiabatic change* and *internal energy*?

A certain mass of an ideal gas is changed from an initial state of pressure p_1 and volume v_1 to a final state p_2, v_2 ($p_2 > p_1$) by two different methods: (i) reversible adiabatic compression from p_1, v_1 direct to p_2, v_2 and (ii) isothermal compression from a volume v_1 to a volume v_2 and then a change of pressure at constant volume v_2 to the final pressure p_2. Show these changes on a p-v diagram. State whether (*a*) the change of internal energy, and (*b*) the work done on the gas is the same, or different, in the two cases, and justify your answers. (*O. and C.*)

6. Distinguish between an *isothermal* and an *adiabatic* compression of a gas. Explain the precautions necessary to ensure that an actual compression approximates to each of these conditions.

A cylinder fitted with a frictionless piston contains 1·0 g of oxygen at a pressure of 760 mmHg and at a temperature of 27 °C. The following operations are performed: (*a*) the oxygen is heated at constant pressure to 127 °C and then (*b*) the oxygen is compressed isothermally to its original volume; and finally (*c*) the oxygen is cooled at constant volume to its original temperature.

Illustrate these changes on a p-V diagram drawn to scale. What is the heat input to the cylinder in stage (*a*)? How much work does the oxygen do in pushing back the piston during this stage? How much work is done on the oxygen in stage (*b*)? How much heat must be extracted from the oxygen in stage (*c*)? (Specific heat capacity of oxygen at constant volume = 0·65 kJ kg^{-1} K^{-1}; density of oxygen at s.t.p. = 1·43 kg m^{-3}; molecular weight of oxygen = 32.) (*C.*)

7. State and discuss the first law of thermodynamics.
Figure 9.8 represents the pressure-volume diagram for a hot-air engine which

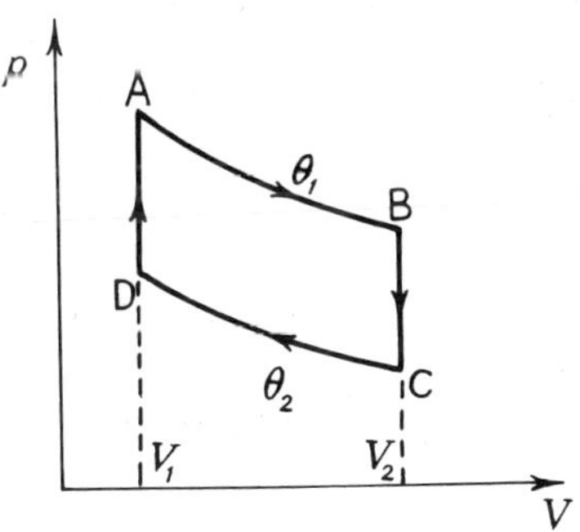

Fig. 9.8

contains as its working substance x moles of air (assumed to behave as an ideal gas) and is imagined to operate cyclically as follows: starting the cycle at A, the engine takes in a quantity of heat energy Q_1 at a high temperature θ_1 K, gives out a quantity of heat energy Q_2 at a lower temperature θ_2 K, and between these two temperatures undergoes appropriate pressure changes at constant volume.

Answer the following questions briefly, giving explanations or reasons: (*a*) Does the internal energy of the air change during the cycle? If so, at what stages and by how much? (*b*) Has the internal energy of the air changed at the end of the cycle? If so, by how much? (*c*) How could you calculate the external work W done by the engine during one complete cycle?

Obtain expressions for Q_1, Q_2 and W in terms of x, θ_1, θ_2, V_1, V_2 and the molar gas constant. (*O.*)

Appendix
Order of Accuracy in Heat Experiments

Errors and order of accuracy

The treatment of errors ranges from the very extensive treatment required in research work to the relatively simple treatment suited to experiments performed by students. In this book we are concerned only with the latter.

There are many errors inherent in a heat experiment. In an experiment to measure the linear expansivity of a metal rod, for example, there will be zero errors and parallax errors in scale readings of lengths, and there will be uncertainty in the contact between the point of the micrometer screw and the end of the bar. In an experiment on specific heat capacity there is uncertainty whether the liquid in the calorimeter has been stirred sufficiently to ensure a uniform temperature throughout. In addition, errors are made in reading temperature and mass from the particular thermometer and balance supplied. All these points need careful consideration by the experimenter; an over-estimation of an error is better practice than an under-estimation.

As illustrated shortly, associated with the final result of an experiment is an *order of accuracy*, which takes into account the effect of all the errors. Thus the measured value of the specific heat capacity of a metal, for example, may be calculated to be 0·50 kJ kg^{-1} K^{-1}. This value may be uncertain by perhaps 0·04 kJ kg^{-1} K^{-1}, so this is an order of accuracy of $(0{\cdot}04/0{\cdot}50) \times 100\%$ or about 8%. In research work it has been possible to determine the charge of an electron to a high degree of accuracy such as 1 part in 5000 or 0·02%; the masses of individual atoms can be found to an order of accuracy of 0·001%.

Calculations of a high order of accuracy can thus be carried out in further investigations on these particles. Unexpected discoveries, or modifications of existing scientific ideas, have also followed from attention to order of accuracy. Rowland's heat experiments, carried out with water at different temperatures, showed a variation which exceeded the estimated order of accuracy. This led to the discovery that the specific heat capacity of water varies with temperature.

Combination of errors

As an illustration of how errors are combined, suppose a reading is estimated to the nearest tenth °C from a suitably calibrated thermometer. A liquid temperature read as 16·4 °C should then be recorded as 16·4 ± 0·1 °C to show the possible error, which covers a temperature range of 16·3—16·5 °C. If the liquid temperature rises and is read as 23·6 °C, it is recorded as 23·6 ± 0·1 °C. Thus

$$\text{temperature rise} = (23{\cdot}6 \pm 0{\cdot}1) - (16{\cdot}4 \pm 0{\cdot}1) = 7{\cdot}2 \pm 0{\cdot}2\,^{\circ}\text{C},$$

since the possible error is ±0·2 °C in a temperature range 7·0 to 7·4 °C. If a more accurately calibrated thermometer is used the error may be less. It should be noted that the errors are *added* when two temperatures are subtracted. This is also the case when two temperatures are added. The percentage error in the temperature rise here is thus

$$\frac{0{\cdot}2}{7{\cdot}2} \times 100\% \quad \text{or} \quad 3\%\ \text{(approx)}$$

Similarly, the mass of a calorimeter is recorded as 34·6 ± 0·1 g if it is weighed to 0·1 g accuracy, or 34·62 ± 0·01 g if it is weighed to 0·01 g accuracy. If the calorimeter is partly filled with water and the whole weighs 88·9 g to 0·1 g accuracy, then, from the difference in weighings,

$$\text{mass of water} = 54{\cdot}3 \pm 0{\cdot}2\,\text{g} \qquad . \ . \ . \ . \ . \quad \text{(i)}$$

If the whole weighs 88·94 g, to 0·01 g accuracy, then

$$\text{mass of water} = 54{\cdot}32 \pm 0{\cdot}02\,\text{g} \qquad . \ . \ . \ . \ . \quad \text{(ii)}$$

For the weighing in (i):

$$\text{percentage error} = \frac{0{\cdot}2}{54{\cdot}3} \times 100\% = 0{\cdot}4\%\ \text{(approx)}$$

For the weighing in (ii):

$$\text{percentage error} = \frac{0{\cdot}02}{54{\cdot}3} \times 100\% = 0{\cdot}04\% \text{ (approx)}$$

Order of accuracy

The final result of any experiment is uncertain to an extent which depends on the errors in each measurement used. We shall consider the effect of the maximum possible error, so that we over-estimate rather than under-estimate.

As a simple illustration, suppose the specific heat capacity c of a solid is calculated from $Q = mct$, where Q is the quantity of heat given to it, m is its mass, and t is the temperature rise. Then

$$c = \frac{Q}{mt}$$

Taking logarithms of both sides,

$$\log c = \log Q - \log m - \log t$$

Differentiating
$$\frac{\delta c}{c} = \frac{\delta Q}{Q} - \frac{\delta m}{m} - \frac{\delta t}{t}$$

where δc, δQ, δm, δt represent small changes such as possible *errors* in c, Q, m, t respectively. The errors are randomly positive or negative, as explained before, and hence

$$\frac{\delta c}{c} = \pm\frac{\delta Q}{Q} \pm \frac{\delta m}{m} \pm \frac{\delta t}{t}$$

The maximum possible error in c is obtained by adding the positive values of the errors in Q, m and t. Hence we write:

$$\frac{\delta c}{c} = \frac{\delta Q}{Q} + \frac{\delta m}{m} + \frac{\delta t}{t} \quad . \quad . \quad . \quad . \quad \text{(i)}$$

The quantity $(\delta c/c) \times 100\%$ is the percentage error in c. Similarly, the terms on the right side of (i), multiplied by 100, represent the percentage errors in Q, m, t respectively.

∴ percentage error in c

= sum of percentage errors in Q, m, t respectively . . (ii)

Suppose $Q = 180{\cdot}0 \pm 0{\cdot}8$ joules, a percentage error of $0{\cdot}4\%$;

$m = 25{\cdot}6 \pm 0{\cdot}2$ g, a percentage error of 0·8%; and $t = 8{\cdot}6 \pm 0{\cdot}2$ °C, a percentage error of 2·3%. Then, from (ii),

$$\text{percentage error in } c = 0{\cdot}4 + 0{\cdot}8 + 2{\cdot}3 = 3{\cdot}5\% \quad . \quad . \quad \text{(iii)}$$

If the errors are temporarily omitted,

$$c = \frac{Q}{mt} = \frac{180}{25{\cdot}6 \times 8{\cdot}6} = 0{\cdot}8176 \text{ by four-figure logarithms.}$$

It is wrong to write the result for c to four places of decimals, since this would imply a percentage error in c of (1/8176) × 100% or 0·012%, whereas the percentage error in c, from (iii), is 3·5%. Now 3·5% of 0·82 is about 0·03. Hence the result should be written

$$c = 0{\cdot}82 \pm 0{\cdot}03 \text{ kJ kg}^{-1} \text{ K}^{-1}$$

The error quoted here is actually 4% of 0·82, but it is convenient to raise the estimated error slightly to a whole number, in keeping with the desire to record the maximum possible limits of error.

Desired accuracy of measurement

From the example just given, the percentage error in the temperature rise t, 2·3%, is relatively much more than the percentage errors in m, 0·8%, and in Q, 0·4%. The percentage error in c, which is found by adding the percentage errors, thus depends to a large extent on the accuracy with which the temperatures are observed. As we have just shown, a temperature rise of 8·6 °C, observed on a thermometer estimated to the nearest tenth °C, is a percentage error of (0·2/8·6) × 100% or 2·3%. If the mass m is weighed to 0·01 g, it might then be 25·64 ± 0·02 g, a percentage error of (0·02/25·64) × 100% or 0·08%. This is very small compared with the error of 2·3%, and would therefore make little difference to the percentage error of c. It is therefore unnecessary to weigh to an accuracy greater than 0·1 g in an experiment of this nature unless a much more accurate thermometer is available.

On the other hand, the final result for the specific latent heat of vaporization l of steam is given by an expression of the form:

$$ml + mc_w(100 - t) = C(t - t_1)$$

where m is the mass of steam condensed in cold water, t is the final water temperature and t_1 is its initial temperature. Thus:

$$l = \frac{C(t - t_1) - mc_w(100 - t)}{m}$$

In this case the percentage error in m in the denominator is added to the percentage error in the numerator to calculate the percentage error in l. If the weighings are carried out to 0·1 g accuracy, the mass of steam, found from the difference in two weighings, will be rather small and may be $3{\cdot}5 \pm 0{\cdot}2$ g. This is a percentage accuracy of about 6%. It is therefore desirable to weigh to an accuracy of 0·01 g before and after steam is added. The mass of steam may then be recorded as $3{\cdot}54 \pm 0{\cdot}02$ g, a percentage accuracy of about 0·6%, and the final result for l is correspondingly more accurate.

A similar case is the measurement of the linear expansivity α of a metal in the form of a long rod or tube (p. 76). Here

$$\alpha = \frac{x}{l \times t} \quad . \quad . \quad . \quad . \quad . \quad . \quad . \quad . \quad \text{(i)}$$

where x is the increase in length of the rod of length l when the temperature rise is t. Taking logarithms of both sides and differentiating, then maximum percentage error in α

$$= \frac{\delta\alpha}{\alpha} \times 100\% = \left(\frac{\delta x}{x} + \frac{\delta l}{l} + \frac{\delta t}{t}\right) \times 100\% \quad . \quad . \quad \text{(ii)}$$

The increase in length x is so small (a little more than a millimetre for a metal rod about a metre long heated through 90 °C), that a micrometer gauge must be used to measure it. If the gauge reads to 0·01 mm, then x may be $1{\cdot}02 \pm 0{\cdot}01$ mm, a percentage error in x of about 1%. On the other hand, if the length l of the rod is about 100 cm, a metre rule used to measure it may give a length of $99{\cdot}6 \pm 0{\cdot}2$ cm, for example. This is a percentage error of about 0·2%. As the percentage error in x is about 1%, it is unnecessary to measure l to a greater accuracy than a metre rule provides. Suppose the temperature rise t is $80{\cdot}0 \pm 0{\cdot}5$ °C, a percentage accuracy of about 0·7%. Then, from (ii),

maximum percentage error in $\alpha = 1\% + 0{\cdot}2\% + 0{\cdot}7\% = 1{\cdot}9\%$

Omitting errors temporarily, and using the above results,

$$\alpha = \frac{0{\cdot}102}{99{\cdot}6 \times 80{\cdot}0} = 12{\cdot}78 \times 10^{-6}\,\text{K}^{-1} \text{ by logarithms}$$

The final result is therefore expressed as

$$\alpha = (12{\cdot}8 \pm 0{\cdot}3) \times 10^{-6}\,\mathrm{K}^{-1}$$

a percentage error of about 2%. In a school laboratory experiment, the result may well be given as $\alpha = (13 \pm 1) \times 10^{-6}\ \mathrm{K}^{-1}$, an error of about 8%.

Revision Papers

Revision Paper 1

1. Describe the structure of a constant-volume gas thermometer. Describe also the method of calibrating it and using it to determine the boiling-point of a salt solution.

Compare this thermometer as a means of measuring temperature with (*a*) a mercury-in-glass thermometer, (*b*) a thermoelectric thermometer. (*L.*)

2. Give an account of a method of determining the specific latent heat of evaporation of water, pointing out the ways in which the method you describe achieves, or fails to achieve, high accuracy.

A 600-watt electric heater is used to raise the temperature of a certain mass of water from room temperature to 80 °C. Alternatively, by passing steam from a boiler into the same initial mass of water at the same initial temperature, the same temperature rise is obtained in the same time. If 16 g of water were being evaporated every minute in the boiler, find the specific latent heat of steam, assuming that there were no heat losses. (*O. and C.*)

3. Explain how the kinetic theory of gases accounts for the pressure of a gas. Obtain an expression for the pressure of a perfect gas, stating clearly what assumptions are made.

Show that the ratio of the velocity of sound in a monatomic gas to the root-mean-square velocity of the molecules is 0·75. Comment on the fact that this ratio is of the order of unity. (*C.*)

4. (*a*) Define *linear expansivity*, and describe how you would determine the linear expansivity of a metal. (*b*) Discuss the effect of a rise in temperature on the time-keeping of a clock with a metal pendulum. (*c*) A vertical column of mercury, 50·0 cm long at 0 °C, is contained in a steel cylinder open at the top. Find in newton metre^{-2} the pressure exerted by the mercury on the base of the cylinder when the temperature is 100 °C. (Take the density of mercury to be 13 600 kg m^{-3}, and linear expansivity of steel to be $1{\cdot}2 \times 10^{-5}$ K^{-1}.) (*O.*)

5. Distinguish between reversible *isothermal* and *adiabatic* changes. An ideal gas is compressed isothermally until its volume is reduced to one-tenth of its initial value, and is then allowed to expand adiabatically to its original volume. Finally the pressure is altered at constant volume until the original state is restored. Represent these changes on a pressure-volume diagram and state whether, on the whole, work has been done on or by the gas. If the initial pressure is 76 cmHg, what is the change of pressure in the last stage of the process? Assume that $\gamma = 1{\cdot}40$. (*L.*)

6. Compare the properties of saturated and unsaturated vapours. By means of diagrams show how the pressure of (*a*) a gas, and (*b*) a vapour, vary with change (i) of volume at constant temperature, and (ii) of temperature at constant volume.

The saturation vapour pressure of ether vapour at 0 °C is 185 mm mercury and at 20 °C is 440 mm. The bulb of a constant-volume gas thermometer contains dry air and sufficient ether for saturation. If the observed pressure in the bulb is 1000 mm at 20 °C, what will it be at 0 °C? (*L.*)

7. Describe how you would determine the thermal conductivity of a good conductor.

The ends of a bar are maintained respectively at 0 °C and 100 °C. Discuss, with the help of diagrams, the temperature distribution along the bar (*a*) when it is well lagged, (*b*) when there is a considerable escape of heat from its side faces.

A copper bar of diameter 1·5 cm has an electric heating coil inserted half-way along its length. The bar is well lagged, except at the ends, which are exposed to the air. When the power supplied to the heater under equilibrium conditions is 12 watts, the steady temperature gradient on each side is 1 °C/cm. Calculate the thermal conductivity of copper.

State briefly the factors that determine the steady temperatures of the exposed faces. (*O.*)

8. Indicate how the concepts of a *black body* and of *black-body radiation* are derived from Prévost's theory of exchanges. Draw curves to illustrate how the energy radiated at different wavelengths varies with the black-body temperature.

The cathode of a diode valve consists of a cylinder 2·0 cm long and 0·10 cm diameter, and is surrounded by a coaxial anode of diameter large compared with that of the cathode. The anode remains at 127 °C when 4 watts is dissipated in heating the cathode. Estimate the temperature of the cathode. List the assumptions you have made in arriving at your estimate. Discuss whether the estimate is realistic. (Stefan's constant = $5{\cdot}74 \times 10^{-8}$ W m^{-2} K^{-4}.) (*C.*)

Revision Paper 2

1. State what is meant by a temperature on the centigrade (Celsius) scale of a platinum resistance thermometer.

Point out the relative merits of (*a*) a platinum resistance thermometer, and (*b*) a thermoelectric thermometer for measuring (i) the rise in temperature of the water flowing through a continuous-flow calorimeter, and (ii) the temperature of a small crystal as it is being heated rapidly. (*N.*)

2. What do you understand by the *specific heat capacity* of a substance? Describe how you would measure the specific heat capacity of a sample of rock, describing the precautions that you would take to obtain an accurate result.

A room is heated during the day by a 1-kW electric fire. The fire is to be replaced by an electric storage heater consisting of a cube of concrete which is heated overnight and is allowed to cool during the day, giving up its heat to the room. Estimate the length of an edge of the cube if the heat it gives out in cooling from 70 °C to 30 °C is the same as that given out by the electric fire in 8 hours.

[Density of concrete = 2700 kg m^{-3}; specific heat capacity of concrete = 0·85 kJ kg^{-1} K^{-1}.] (*O. and C.*)

3. State the laws of gases usually associated with the names of Boyle, Charles, Dalton, and Graham. Two gas containers with volumes of 100 cm^3 and 1000 cm^3

respectively are connected by a tube of negligible volume, and contain air at a pressure of 1000 mm of mercury. If the temperature of both vessels is originally 0 °C, how much air will pass through the connecting tube when the temperature of the smaller is raised to 100 °C? Give your answer in cm^3 measured at 0 °C and 760 mm of mercury. (*L.*)

4. Describe how you would determine, in your school laboratory, the coefficient of linear expansion of a metal supplied in the form of a rod or tube about 60 cm long. Indicate the likely experimental error in the measurements made and hence deduce the maximum possible percentage error in the final result.

A silica rod has a diameter of 0·75 cm at 20 °C and just fits into an aluminium tube at this temperature. Find the temperature at which the cross-sectional area of the space between the rod and the tube is 0·50 mm^2. The coefficients of linear expansion of silica and aluminium are $0{\cdot}50 \times 10^{-6}$ and $25{\cdot}5 \times 10^{-6}$ K^{-1} respectively. (*N.*)

5. "The two specific heat capacities in kJ kg^{-1} K^{-1} units for argon are 0·521 and 0·313 and for air are 1·012 and 0·722." Explain these statements and discuss their significance in relation to (*a*) the atomicity of the molecules of the two gases, (*b*) the relative values of the adiabatic elasticities of argon and air.

Describe an experiment to verify *one* of the above values of specific heat capacities. (*L.*)

6. Without deriving any formulae, use the kinetic theory of gases to explain (*a*) how a gas exerts a pressure, (*b*) why the temperature of a gas rises when the gas is compressed, (*c*) what happens when a quantity of liquid is introduced into a closed vessel. How are the differences in the behaviour of real and ideal gases explained by the kinetic theory? If there are $2{\cdot}7 \times 10^{19}$ molecules in a cubic centimetre of gas at 0 °C and 760 mm mercury pressure, what is the number per cubic centimetre (i) at 0 °C and 10^{-6} mm pressure, (ii) at 39 °C and 10^{-6} mm pressure? (*N.*)

7. Define *coefficient of thermal conductivity*, and describe how you would determine the conductivity of a bad conductor, such as cork.

A thin-walled metal hot-water tank, of effective surface area 6 square metres, is covered with a layer of insulation 3 cm thick. The coefficient of thermal conductivity of the insulating material is 10^{-1} W m^{-1} K^{-1}. Find the power which must be supplied by an electric immersion heater to maintain the water at 60 °C when the outer face of the insulation is at 30 °C.

Taking the air temperature to be constant at 15 °C, and assuming that the temperature difference between the outer face of the lagging and the air is proportional to the power supplied, find the temperature of the water in the tank when the power supply is raised to 867 watts. (*O.*)

8. Discuss the nature of the processes by which a hot body may lose heat to the surroundings.

A blackened strip of area 0·20 cm^2 is placed at a distance of 200 cm from a white-hot iron sphere of diameter 1·0 cm, so that the radiation falls normally on the strip. The radiation causes the temperature, and hence the resistance, of the platinum to increase. It is found that the same increase in resistance can be produced under similar conditions, but in the absence of radiation, when a current of 3·0 mA is passed through the platinum strip, the potential difference between its ends being 24 mV. Estimate the temperature of the iron sphere. (Stefan's constant = $5{\cdot}7 \times 10^{-8}$ W m^{-2} K^{-4}.) (*C.*)

Revision Paper 3

1. Three types of thermometer in common use are based on (*a*) the expansion of a fluid, (*b*) the production of an electromotive force, (*c*) the variation of electrical resistance. Describe briefly one example of each and the way in which it is used. In each case, state how a value of the temperature on a centigrade scale is deduced from the quantities actually measured.

If all three of the thermometers you have described were used to measure the temperature of the same object, would they give the same result? Give reasons for your answer. (*O. and C.*)

2. 300 g of a certain metal of density about 10 g cm^{-3} is available in the form of a coarse powder, together with a calorimeter of heat capacity about 33·6 J K^{-1} and volume about 160 cm^3, and a 50 °C thermometer reading to $\frac{1}{5}$ deg.

Using this and other necessary apparatus, how would you verify, by the method of mixtures, that the specific heat of the metal is 0·13 kJ kg^{-1} K^{-1}?

In the experiment you describe why is it (*a*) unnecessary to apply a correction for heat exchange with the surroundings, (*b*) necessary to decide on a suitable maximum temperature of the mixture? How would you ensure that such a temperature is realized? (*N.*)

3. What do you understand by the term *ideal gas*? Discuss the experimental evidence that leads to the ideal gas equation $pV = RT$.

The bulb of a simple constant-volume air thermometer has a volume of 120 cm^3, and it is connected to the manometer by tubing of volume 10 cm^3, which remains at room temperature (15 °C) throughout an experiment. When the bulb is immersed in an ice-water mixture at 0 °C, the gas in the bulb is at a pressure of 880 mm mercury. To what pressure will it be subjected when the bulb is immersed in steam at 100 °C? (*O.*)

4. Describe and explain a method of measuring the absolute coefficient of expansion of a liquid. A mercury barometer with brass scale reads 765·3 mm at 20 °C. What would the reading be at 0 °C, the atmospheric pressure remaining the same? [Cubic expansivity of mercury = 181×10^{-6} K^{-1}; linear expansivity of brass = 19×10^{-6} K^{-1}.] (*L.*)

5. Obtain from first principles an expression for the temperature change of an ideal gas that undergoes a reversible adiabatic change.

In terms of volumes, the compression ratio of an internal-combustion engine is 8:1. If the fuel/air mixture is drawn into the cylinder at a uniform temperature of 50 °C, find the temperature to which it is raised. (Take the ratio of the two principal specific heat capacities of the mixture to be 1·36.) (*O.*)

6. What do you understand by an *ideal gas*? Discuss the extent to which the behaviour of (*a*) real gases, and (*b*) saturated vapours can be represented by relations derived for an ideal gas.

Describe an experiment to investigate the relation between the pressure and temperature of a sample of air maintained at constant volume over the temperature range from 0 °C to 100 °C.

A tube of length 40 cm and uniform area of cross-section 1·0 cm^2, held vertically, is pushed downwards into a trough of mercury until its top is 15·0 cm above the mercury level. An airtight cap is placed over the top end of the tube and the tube is very slowly raised, keeping its axis vertical, until the bottom end of the tube is 1·0 cm below the level of the mercury in the trough. What is the new height of mercury in the tube, assuming that the temperature of the air in the tube remains constant? Take the atmospheric pressure as 760 mmHg, and neglect surface tension effects. (*C.*)

7. Explain what is meant by the *coefficient of thermal conductivity* of a substance, and describe an experiment to determine this quantity for copper.

A copper kettle has a circular base of radius 10 cm and thickness 3·0 mm. The upper surface of the base is covered with a uniform layer of scale 1·0 mm thick. The kettle contains water which is brought to the boil over an electric heater. In the steady state 5·0 g of steam is produced each minute. What is the temperature of the lower surface of the base, assuming that conduction of heat up the sides of the kettle can be neglected? (Values of thermal conductivity: copper, 382 W m^{-1} K^{-1}; scale, 1·34 W m^{-1} K^{-1}. Specific latent heat of vaporization of water = 2268 kJ kg^{-1}.) (*C.*)

8. Give an account of the transmission of heat energy by radiation.

Describe any way of making a close approximation to a perfectly black body which would be useful for thermal radiation experiments over the wavelength range between about 10^{-4} cm to 10^{-3} cm. Explain how the distribution of energy between the various wavelengths in the spectrum of a black body, and also the total radiation it emits per unit area per second, depend on the absolute temperature.

A heating panel of area 2·5 m^2 is supplied with energy at a steady rate of 1·5 kilowatt. Calculate the surface temperature of the panel when the surroundings are at 20 °C, assuming that heat exchange takes place only by radiation and that both the panel and the surroundings behave as perfectly black bodies. (Take Stefan's constant to be $5{\cdot}7 \times 10^{-8}$ W m^{-2} K^{-4}.) (*O.*)

Revision Paper 4

1. Explain the principle of a constant-volume gas thermometer and describe a simple instrument suitable for measurements in the range 0° to 100 °C. What factors determine (*a*) the sensitivity and (*b*) the accuracy of the instrument you describe?

A certain gas thermometer has a bulb of volume 50 cm^3 connected by a capillary tube of negligible volume to a pressure gauge of volume 5·0 cm^3. When the bulb is immersed in a mixture of ice and water at 0 °C, with the pressure gauge at room temperature (17 °C), the gas pressure is 700 mmHg. What will be the pressure when the bulb is raised to a temperature of 50 °C if the gauge is maintained at room temperature? You may assume that the gas is ideal and that the expansion of the bulb can be neglected. (*O. and C.*)

2. Give an account of an electrical method of finding the specific latent heat of vaporization of a liquid boiling at about 60 °C. Point out any causes of inaccuracy and explain how to reduce their effect.

Ice at 0 °C is added to 200 g of water initially at 70 °C in a vacuum flask. When 50 g of ice has been added and has all melted, the temperature of the flask and contents is 40 °C. When a further 80 g of ice has been added and has all melted, the temperature of the whole becomes 10 °C. Calculate the specific latent heat of fusion of ice, neglecting any heat lost to the surroundings.

In the above experiment the flask is well shaken before taking each temperature reading. Why is this necessary? (*C.*)

3. What are the main assumptions of the *kinetic theory of gases*? Explain fully how the ideal gas equation $pV = RT$ is derived from the kinetic theory.

The density of air is 1·3 kg m^{-3} at 0 °C and 10^5 N m^{-2} pressure. Calculate the r.m.s. velocity of air molecules at 27 °C and 10^5 N m^{-2} pressure.

4. Define the *cubic expansivity* of a liquid. How does the density of a liquid depend on the temperature?

Describe how you would determine the absolute coefficient of expansion of mercury.

It is required to design a mercury-in-glass thermometer for use over the range 0–200 °C using the following data: absolute cubic expansivity of mercury, $1{\cdot}84 \times 10^{-4}$ K^{-1}; linear expansivity of glass = 8×10^{-6} K^{-1}; length of scale, 25 cm; diameter of cylindrical bulb, 3 mm. Suggest suitable figures for the bore of the capillary tube and the length of the bulb. (*O.*)

5. Explain what is meant by a reversible adiabatic change. Use the concepts of simple kinetic theory to explain why the temperature of a gas enclosed in a cylinder by a piston will rise while the piston moves so as to reduce the volume.

Describe how the ratio γ of the two principal specific heat capacities of a gas may be determined by a method involving a single adiabatic expansion, such as that of Clément and Désormes. (*O.*)

6. Distinguish between an *isothermal* and an *adiabatic* compression of a gas. Explain the precautions necessary to ensure that an actual compression approximates to each of these conditions.

Sketch curves to show the relationship between pressure and volume, determined under isothermal conditions at a number of temperatures between about 10 °C and 40 °C for (*a*) carbon dioxide, (*b*) hydrogen. Explain the form of the curves.

Give a brief description of a method for liquefying hydrogen, emphasizing the physical principles involved rather than the technical details. (*C.*)

7. Define the coefficient of thermal conductivity. Describe a method of measuring the thermal conductivity of a solid which is a bad conductor of heat, and explain carefully why it is not possible to employ the same method as that used for a good conductor.

A pond is 40 cm deep. If the air temperature above the water is $-5{\cdot}0$ °C and the temperature of the water at the bottom of the pond is maintained at 4·0 °C, find the thickness of the ice that will eventually be formed. (Thermal conductivity of ice $= 2{\cdot}3\ \mathrm{W\,m^{-1}\,K^{-1}}$, of water $= 0{\cdot}56\ \mathrm{W\,m^{-1}\,K^{-1}}$.) (*C.*)

8. State the factors which determine the rate at which a body loses heat by radiation. Discuss one method by which the temperature of a radiating body can be determined, and mention the assumptions involved.

A flat surface disc 2 cm thick is held so that the sun's radiation falls normally upon one face. Given that Stefan's constant is $5{\cdot}7 \times 10^{-8}\ \mathrm{W\,m^{-2}\,K^{-4}}$, the radius of the sun is 7×10^5 km, the mean distance from earth to sun is $1{\cdot}5 \times 10^8$ km, and assuming that the sun radiates as a black body at a temperature of 6000 K, calculate in watts per cm^2, the rate of incidence of solar energy on the disc.

If the steady temperatures reached by the two faces of the disc are 110 °C and 86 °C while the surroundings are at 7 °C, calculate the thermal conductivity of the material of the disc on the supposition that the surface facing the sun can be treated as a perfectly black body and that no heat escapes from the edge. (*O.*)

Answers to Exercises

Exercise 1 (p. 10)

1. (i) 292·1; (ii) 15·2 K
2. (i) 267·7; (ii) 311·4 K
3. (i) −100 °C; (ii) 150 °C
8. 50·4 °C
9. 68 °C, −272 °C
11. 385 °C

Exercise 2 (p. 36)

1. (i) 25 °C; (ii) 42 kJ
2. 0·08 kJ K^{-1}, 30 °C
3. 1/3 °C
4. 42 kJ, 14 min
5. 0·8 kJ kg^{-1} K^{-1}
6. (i) 656 kJ; (ii) 7·6 kJ
7. (i) 326 kJ kg^{-1}; (ii) 5·25 g
8. 164 g
9. 1·4 kJ kg^{-1} K^{-1}
10. 14·8
11. 3·05 kJ kg^{-1} K^{-1}
12. 4170 J kg^{-1} K^{-1}, ±5%
14. 2·2 kJ kg—[1] K^{-1}
15. 0·74 kJ kg^{-1} K^{-1}, 30·8 min
16. 0·6, 1·1, 1·9 kJ kg^{-1} K^{-1}
17. 351 kJ kg^{-1}
18. 74 km h^{-1}
19. 904 m
20. 8·8 g
21. 417 kJ kg^{-1}, 6·7 W

Exercise 3 (p. 69)

1. Charles
2. molar gas constant; 8·3
3. $p\,.\,\delta V$; external
4. rate, momentum per unit area
5. mean square, mean
6. $3{\cdot}66 \times 10^{-3}$ K^{-1}, 478 cm^3
7. (i) 78·5; (ii) 73·2 cmHg
8. 131 °C
9. 241 cm^3
10. (i) 0·29; (ii) 2·9 kJ kg^{-1} K^{-1}; 0·99 g
11. 12 J
12. 100 °C
13. 3·75 atm
14. 0·014 m^3
15. (i) 2·08 kJ kg^{-1} K^{-1}; (ii) 842 mmHg
16. 878 mmHg
17. 1840, 2150 m s^{-1}
19. $1{\cdot}5 \times 10^5$ J m^{-3}
20. 597 m s^{-1}
22. 0·21 mmHg
23. 1305 m s^{-1}, 0 °C
24. (*a*) $1{\cdot}7 \times 10^{16}$ mols; (*b*) 3×10^{-5} cm;
(*c*) $1{\cdot}9 \times 10^3$ m s^{-1}; (*d*) about 10^{18} cm^{-2} s^{-1}

Exercise 4 (p. 95)

1. 434 °C
2. 300 N
3. 270 °C
4. 4s
5. (*a*) 95·2 cm; (*b*) 0·53 cm
6. 0·436, 0·444 cm^3
7. 762·4 mmHg
8. $3{\cdot}8 \times 10^{-4}$ cm^2
9. $8{\cdot}5 \times 10^{-6}$ K^{-1}
10. 79 °C
11. 20·5 cm
12. 62·4 g
13. 12×10^{-6} K^{-1}
14. 964 g

Exercise 5 (p. 119)

1. independent
2. R
3. intermolecular attraction
4. c_V; $\frac{1}{2}RT$
5. 0·71 kJ kg^{-1} K^{-1}
6. 64 cmHg
7. 30·4 J. (i) gain (ii) gain
8. 163·5 cmHg, 80 °C
9. 10·3 kJ kg^{-1} K^{-1}
10. 4·17 kJ kg^{-1} K^{-1}
12. 189 K
13. 56·8 cmHg; 227 K
14. 164 J
15. 0·725 kJ kg^{-1} K^{-1}
17. (*a*) 586 K; (*b*) 101·2 J; (*c*) 354 J
18. 85 °C
19. 0·53 kJ kg^{-1} K^{-1}; 1/3

Exercise 6 (p. 149)

1. 855 mmHg
3. 44·5 mmHg
7. 91·7 torr
9. 707 mmHg
11. 78 %
18. (i) P_1; (ii) P_2
21. 146·5 mmHg

Exercise 7 (p. 175)

1. (i) 1700 K m^{-1}; (ii) 500 K m^{-1}
2. 192×10^4 J
3. 8×10^4 J
7. 355 J
9. (*a*) 90 °C; (*b*) 58·4 W
12. 1·7 kg
13. (*a*) 995; (*b*) 159 K m^{-1}
14. 42 W m^{-1} K^{-1}, 0·12 W m^{-1} K^{-1}
15. $1{\cdot}4 \times 10^5$ K m^{-1}
16. 41 : 1
18. $I^2\rho/4\pi^2kR^2$

Exercise 8 (p. 196)

2. (i) 1·5 : 1; (ii) 2·2 : 1
3. 5470 J
5. 6×10^{-15} W m^{-2} K^{-4}; 1·7 (approx)
6. 2014 K
7. 1780 K
8. 15·6 W
10. 5450 °C
11. 5490 °C
13. $5{\cdot}6 \times 10^7$ J
14. $5{\cdot}7 \times 10^{-8}$ W m^{-2} K^{-4}
15. 22 W, 10 %
16. 2140 K
17. 0·9 : 1

Exercise 9 (p. 211)

2. 6·26, 2·09 atm; 353 °C
3. $c_V(p_2V_2 - p_1V_2)/R$
4. (*a*) 10^5 N m^{-2}; (*b*) 75 K; (*c*) 3600 J; 2400 K
5. (*a*) same; (*b*) different
6. (*a*) 92 J, 26 J; (*b*) 30 J; (*c*) 67 J
7. (*a*) $\pm 1{\cdot}5xR(\theta_2 - \theta_1)$ along BC, DA; (*b*) no; (*c*) area ABCD;
$Q_1 = xR\theta_1 \log_e (V_2/V_1)$, $Q_2 = xR\theta_2 \log_e (V_2/V_1)$, $W = xR(\theta_1 - \theta_2)\log_e (V_2/V_1)$

Revision Paper 1 (p. 220)

2. 2230 kJ kg^{-1}
4. 67 840 N m^{-2}
5. on the gas; 46 cmHg
6. 707 mmHg
7. 340 W m^{-1} K^{-1}
8. 1030 K

Revision Paper 2 (p. 221)

2. 0·68 m
3. 33 cm^3
4. 246 °C
6. (i) 3·6 × 10^{11}; (ii) 3·1 × 10^{11}
7. 400 W, 80 °C
8. 1780 K

Revision Paper 3 (p. 222)

3. 1170 mmHg
4. 762·8 mm
5. 410 °C
6. 19 cm
7. 104·5 °C
8. 93 °C

Revision Paper 4 (p. 224)

1. 815 mmHg
2. 378 kJ kg^{-1}
3. 500 m s^{-1}
4. bulb length, 10–20 mm; bore, 0·13–0·16 mm
7. 33·5 cm
8. 0·16 W cm^{-2}, 0·61 W m^{-2} K^{-4}

Index